Chemistry HL
FOR THE IB DIPLOMA

Cameron Lumsden

PEAK

Published by:
Peak Study Resources Ltd
1 & 3 Kings Meadow
Oxford OX2 0DP
UK

www.peakib.com

Chemistry HL: Study & Revision Guide for the IB Diploma

ISBN 978-1-913433-24-6

© Cameron Lumsden 2015–2020

Cameron Lumsden has asserted his right under the Copyright, Design and Patents Act 1988 to be identified as the author of this work.

All rights reserved. No part of this publication may be reproduced, stored in a retrieval system, or transmitted in any form or by any means, without the prior permission of the publishers.

*PHOTOCOPYING ANY PAGES FROM THIS PUBLICATION,
EXCEPT UNDER LICENCE, IS PROHIBITED*

Peak Study & Revision Guides for the IB Diploma have been developed independently of the International Baccalaureate Organization (IBO). 'International Baccalaureate' and 'IB' are registered trademarks of the IBO.

Orders: books may be ordered directly through the publisher's website www.peakib.com, or to enquire about local stockists please contact us at books@peakib.com (schools@peakib.com for educational establishments).

Printed and bound in the UK by CPI Group (UK) Ltd, Croydon CR0 4YY

www.cpibooks.co.uk

Table of Contents

Quantitative Chemistry 7

Atomic Structure 29

Periodicity 43

Bonding 59

Energetics 83

Kinetics 103

Equilibrium 115

Acids & Bases 125

Oxidation & Reduction 141

Organic Chemistry 151

Measurement & Data Processing 185

Acknowledgements

Thanks to Dr. Peter Morgan of Taipai American School, and Martin Bluemel of Taunton School in Somerset for their gracious assistance in proofreading the original text.

The purpose of this book

The purpose of this book is to highlight the types and styles of questions that are likely to appear on your IB Chemistry exam. There are Learning Check Questions designed to focus on a specific skill, and there are end of chapter questions that mimic the type of questions that IB Chemistry exams contain.

This book also highlights spots where students can often go wrong, where they get "trapped" by a tricky question, and how they should present their answers.

Who this book is for...

This book is intended for students who have already been through the course material with a teacher and other resources. No text book can replace the ability of a teacher to guide student understanding.

How to use this book.

In many places in this text there is space for you to fill in information, but there is not a lot of blank space. You should keep a separate set of revision notes that are a hybrid of this book, your notes, your text etc.

Other resources

While studying it is important that you also have the following materials..

The Syllabus: You should use this as a check list, and understand what you know, and more importantly, what you don't know.

The Data Booklet: To date, the data booklet has not been used heavily by IBO during the exams, but you should be familiar with what is in it.

Past Papers: All teachers and IB Co-ordinators have access to past papers. Do as many as you can to get yourself familiar with the style and type of questions.

Your own course notes. - Hopefully you have maintained an accurate record of the material that you have covered in your classes. You should compare the way your teacher presents information, the way your textbook presents information, and the way information is presented here. They should all be effectively the same! - However, perhaps one makes more sense to you. Compare and learn!

How to study

The best way to study for retention of information is to put yourself on a repeating cycle.

During your course, you have to do the daily work that the teacher assigns, but if you can, you should also regularly...

- Review what you did yesterday (last class)
- Review what you did last week
- Review what you did last month.

A tall order, but very effective if you can find the time - really you should make time - it's well invested as it will save you trying to "re-learn" everything.

Definitions

All definitions in this text are derived from the IUPAC Goldbook; http://goldbook.iupac.org.

Exam Review Time

You no doubt have six exams that you need to study for, but probably not all of them have the same priority.

Whatever the subject, you are best trying to learn slowly and regularly. If you have purchased this book in the Spring before your exams, you should make a plan to work on each unit of the course. You will probably need at least a week for each topic to...

- make notes
- learn definitions
- make flash cards (great if you take a bus to school)
- practice specific problem types
- do past paper questions

A warning! Cramming doesn't work. You either are not effective in learning material, you learn it superficially, or you run out of time to generate real understanding.

Slow and steady wins the race!

The Exams you face...

Paper 1 - 1:00 hour: 45 Multiple Choice questions (4 choices). No penalty for guessing.
NO CALCULATOR, NO DATABOOK

Paper 2 - 2:15 hours: Short answer and extended response - 95 points
Calculator and Databook allowed.

Paper 3 - 1:30 hours: Part 2: Questions on the practical experiences required in the syllabus.
Part 1: Questions on the option topic that you have studied. Short answer only.

Paper 1 & 2 are written "back to back" on the afternoon of the given day. Meaning that you will be in the exam room for a possible 3:30 hours - by the time you get seated etc. Get comfortable!

Paper 2 & 3 are allowed 5 minutes reading time.

Your Grade

Paper 1   20%
Paper 2   36%
Paper 3   24%
Individual Investigation   20%

Note: The "Core" of the course - Paper 1&2 count for 56% of you final grade.

While different papers are weighted differently, it works out that each raw point on any given component is worth about 0.5% of your final grade.

Typical Overall HL Grade Boundaries

These vary from year to year but the following table gives you a **rough** idea.

Final Grades are calculated based on your Total RAW scores of all the components weighted together, not the 1-7 scale - 39/40 is better than 38/40 even though they are both "7" - The "7" is for guidance only.

1	2	3	4	5	6	7
0–19%	20% – 34%	35% – 47%	48% – 59%	60% – 69%	70 – 79%	80 – 100%

The Periodic Table of Elements

Atomic Number
Symbol
Relative mass

1 H 1.01																	2 He 4.00
3 Li 6.94	4 Be 9.01											5 B 10.81	6 C 12.01	7 N 14.01	8 O 16.00	9 F 19.00	10 Ne 20.18
11 Na 22.99	12 Mg 24.31											13 Al 26.98	14 Si 28.09	15 P 30.97	16 S 32.07	17 Cl 35.45	18 Ar 39.95
19 K 39.10	20 Ca 40.08	21 Sc 44.96	22 Ti 47.90	23 V 50.94	24 Cr 52.00	25 Mn 54.94	26 Fe 55.85	27 Co 58.93	28 Ni 58.71	29 Cu 63.55	30 Zn 65.37	31 Ga 69.72	32 Ge 72.59	33 As 74.92	34 Se 78.96	35 Br 79.90	36 Kr 83.80
37 Rb 85.47	38 Sr 87.62	39 Y 88.91	40 Zr 91.22	41 Nb 92.91	42 Mo 95.94	43 Tc 98.91	44 Ru 101.07	45 Rh 102.91	46 Pd 106.42	47 Ag 107.87	48 Cd 112.40	49 In 114.82	50 Sn 118.69	51 Sb 121.75	52 Te 127.60	53 I 126.90	54 Xe 131.30
55 Cs 132.91	56 Ba 137.33	57 La* 138.91	72 Hf 178.49	73 Ta 180.95	74 W 183.85	75 Re 186.21	76 Os 190.21	77 Ir 192.22	78 Pt 195.09	79 Au 196.97	80 Hg 200.59	81 Tl 204.37	82 Pb 207.19	83 Bi 208.98	84 Po (210)	85 At (210)	86 Rn (222)
87 Fr (223)	88 Ra (226)	89 Ac** (227)	104 Rf (267)	105 Db (268)	106 Sg (269)	107 Bh (270)	108 Hs (269)	109 Mt (278)	110 Ds (281)	111 Rg (281)	112 Cn (285)	113 Uut (286)	114 Fl (289)	115 Uup (288)	116 Lv (293)	117 Uus (294)	118 Uuo (294)

*	58 Ce 140.12	59 Pr 140.91	60 Nd 144.24	61 Pm (145)	62 Sm 150.36	63 Eu 151.96	64 Gd 157.25	65 Tb 158.93	66 Dy 162.50	67 Ho 154.93	68 Er 167.26	69 Tm 168.93	70 Yb 173.05	71 Lu 174.97
**	90 Th 232.04	91 Pa 231.04	92 U 238.03	93 Np (237)	94 Pu (244)	95 Am (243)	96 Cm (247)	97 Bk (247)	98 Cf (251)	99 Es (252)	100 Fm (257)	101 Md (258)	102 No (259)	103 Lr (262)

Chapter 1

Quantitative Chemistry

In this chapter...

8	Introduction
8	Mole Concept
9	Units
9	Mole – Particle Conversions
10	Molar Mass
10	Mole – Mass Conversions
11	Balancing Equations
12	Mole Relationships in a Chemical Reaction
13	Mass Relationships in a Chemical Reaction
14	Limiting Reactant
15	Determination of Formulae - Gravimetric Analysis
16	Mixtures & Solutions
17	Making solutions and determining solute mass
17	Kinetic Molecular Theory
17	Pressure
18	Boyle's Law
18	Pressure Law
19	Charles' Law
19	The Combined Gas Law
20	Avogadro's Law
20	Avogadro's Law of Combining Volumes
21	Ideal Gas Law
22	Yield: Theoretical, Experimental and Percentage
22	Empirical & Molecular Formulae
23	Calculations from percentage information
24	Calculations from Empirical Data
25	Compounds containing oxygen
26	Molecular Formula
28	Summary Questions

Introduction

Jeremias Benjamin Richter (1762-1807) was the first to lay down the principles of stoichiometry. In 1792 he wrote:

> *"Die Stöchyometrie (Stöchyometria) ist die Wissenschaft die Quantitativen oder Massenverhältnisse zu messen, in welchen die chymischen Elemente gegen einander stehen."*

> *"Stoichiometry is the science of measuring the quantitative proportions or mass ratios in which chemical elements stand to one another."*

Or more simply put, everything in this unit is a simple ratio. As long as you know one quantity, you can figure out the others by applying a ratio.

Special Note on Units

Be clear and organized. You may have many numbers floating around on your page. It's very helpful if, after solving a step in a problem, you write " *n=5.5 mol of H_2*", or n(H_2)=5.5 mol not just "*5.5*" **always include the units** and the species, to remind yourself of what you have solved for.

Special Note on Sig Figs.

Throughout this unit (and all others), you should be obeying the rules of significant figures. Significant Figures are covered in Chapter 11.

Worked examples in this chapters show intermediate answers which are rounded to the appropriate number of significant figures, However, the calculator answer is retained for further steps.

You should retain the calculator value in your calculations, and only round the final answer.

Mole Concept

Chemists work on a particle basis – one particle of this reacts with three particles of that to produce two particles of something else. It doesn't matter at the moment what those particles actually are – they could be atoms, diatomic molecules, larger molecules, ions, electrons, etc. The problem is that it takes an awful lot of particles to get an amount that can be manipulated on the "human scale" – as opposed to the "atomic scale".

So knowing what the atomic masses were in terms of protons, neutrons (and electrons), it was decided that we would use the same value of atomic mass and scale it up to the "gram-equivalent mass". As long as all elements were scaled up the same amount, the relative mass would be the same.

It turns out that the scale factor is rather large (because atoms are rather small!) – it is 6.02×10^{23}.

Definition

One mole contains the same number of particles as 12.000 g of ^{12}C.
12.000 g of ^{12}C contains 6.02×10^{23} carbon atoms.

Now, you have to be careful of what types of particles you are talking about.

Example

1.0 mol of H_2O contains...	(Notes)
6.02×10^{23} molecules of water	also called "Avogadro's Number"
6.02×10^{23} oxygen atoms	the same number of oxygen atoms
1.2×10^{24} atoms of hydrogen	double the number of hydrogen atoms
1.8×10^{24} total atoms	triple the number of total atoms

Units

Just to put moles into perspective - moles are the same as any other unit. Moles measure the "amount of substance", which fundamentally means the **number** of particles - not the qualities of those particles, like mass or volume.

Quantity	Unit	Symbol	Example
Mass	gram*	g	m= 5.0 g
Length	meter	m	l= 0.025 m
Time	second	s	t= 35.5 s
Temperature	Kelvin	K	T= 298 K
	Degrees Celsius	°C	T= 25.2°C
Amount of substance	Mole	mol	n= 3.5×10^{-2} mol

Figure 1.1: SI Units

"Amount" always means moles

Note: We use "n" for *n*umber of moles.

* the actual SI unit for mass is the kilogram, but chemists use grams

Mole – Particle Conversions

There are lots of different ways to remember this... The easiest is to remember the idea- it takes 6.02×10^{23} particles to make 1.0 mol.

$$\text{Number of Particles} = \text{number of moles} \times 6.02 \times 10^{23}$$

Expression

Remember: it doesn't matter what mass the particles have – you are just converting the number – it's like converting single eggs to dozens of eggs and vice versa – this is easy, because the value for "dozen" is always 12.

Determine the number of particles in the following amounts.

a) 2.0 mol of neon
b) 0.125 mol of zoopers
c) 1.5 mol of widgets
d) 1.25×10^{-3} mol of ions
e) 12.5 μmol of electrons*
f) 4.55×10^{-9} mol of particles
g) 5.3 mol of gold atoms
h) 1.85×10^{-2} mol of things

* μ = 10^{-6}

1.1 Learning Check

Determine the number of moles in the following amounts.

a) 6.02×10^{25}
b) 2.26×10^{23}
c) 3.31×10^{24}
d) 7.53×10^{20}
e) 3.46×10^{18}
f) 5.27×10^{19}
g) 2.56×10^{20}
h) 1.05×10^{24}

1.2 Learning Check

Molar Mass

Molar mass is the mass of one mole of a substance, and is the "gram-equivalent mass" of the relative atomic mass on the periodic table.

The easiest way to think about this is that the mass numbers on the periodic table can be thought of as follows...

1 atom of oxygen	= 16 atomic mass units (8 p + 8 n)
6.02×10^{23} atoms of oxygen	= 16.00 g of oxygen atoms
1 mol of oxygen atoms	= 16.00 g of oxygen atoms

Notice the last two are the same!

The numbers on the periodic table are for ONE MOLE or ONE ATOM, so everything is just a multiple of it.

Example

$$Al_2(SO_4)_3 = 2 \times Al = 2 \times 26.98 \text{ g mol}^{-1} = 53.96 \text{ g mol}^{-1}$$
$$+ 3 \times S = 3 \times 32.06 \text{ g mol}^{-1} = 96.18 \text{ g mol}^{-1}$$
$$+ 12 \times O = 12 \times 16.00 \text{ g mol}^{-1} = 192.00 \text{ g mol}^{-1}$$
$$Al_2(SO_4)_3 = 342.14 \text{ g mol}^{-1}$$

HINT: Your calculator knows the order of operations, try doing it all on one line!

1.3 Learning Check

Calculate the molar mass of the following formulas. (To two decimal places)
- a) NaCl
- b) Na_2CO_3
- c) NH_4NO_3
- d) $CaCl_2$
- e) $Ca_3(PO_4)_2$
- f) $MgSO_4 \cdot 7H_2O$
- g) Fe_2O_3
- h) $(NH_4)_3PO_4$
- i) CH_3COOH

NOTE: in (f) the compound is a hydrate, and contains 7 moles of water for each mole of ionic salt.

Mole – Mass Conversions

To find the mass of a given number of moles, we simply need to multiply by how many moles we have, and we know the mass of one mole from the periodic table.

Expression

$$\boxed{\text{mass} = \text{number of moles} \times \text{molar mass}}$$

in terms of units...

Expression

$$\boxed{\text{grams} = \text{moles} \times \frac{\text{grams}}{\text{mole}}}$$

1.4 Learning Check

Calculate the mass of the following.
- a) 1.50 mol of LiCl
- b) 2.25 mol of $Ca(OH)_2$
- c) 0.165 mol of $CuCl_2$
- d) 7.50×10^{-6} mol of Al_2O_3
- e) 5.00×10^{-3} mol of Na_2CO_3
- f) 8.75×10^{-9} mol of PbI_2
- g) 0.750 mol of $MgSO_4$
- h) 3.75×10^{-6} mol of $AgNO_3$
- i) 2.500 mol of H_2SO_4

1.5 Learning Check

Calculate the amount (number of moles) of the following.
- a) 10.0 g of $CuSO_4$
- b) 10.0 g of $MgSO_4$
- c) 10.0 g of Na_2SO_4
- d) 0.550 g of NaOH
- e) 6.25 g of $BaCl_2$
- f) 55.8 g of ZnO
- g) 25.75 g of $SnCl_4$
- h) 5.05 g of $HgCl_2$
- i) 1.00 g of $Na_2B_4O_7$

Balancing Equations

A chemical reaction can never create or destroy matter (atoms); it can only rearrange them. Therefore, we must have the same number of atoms of each type on each side of the equation.

Balancing equations means that you multiply each (correct) formula by a coefficient so that you obtain equal numbers of each type of atom on each side of the equation.

Some people say there are rules for balancing equations, but they are more like guidelines. The best guideline is...

Leave uncombined elements until the end – e.g. $O_2(g)$, $H_2(g)$, $K(s)$, etc., because when you change their coefficient, you don't mess up anything else!

At the moment, we don't want any fractions – so you may have double all coefficients in order to clear the fraction.

Later we might need to use fractions because our equation must obey a definition (usually in order to produce only one mole of product) – See Energetics - Heat of Combustion and Heat of Formation.

Tip

Balance the following equations.

a) __K + __H_2O → __H_2 + __KOH

b) __CuO + __NH_3 → __H_2O + __N_2 + __Cu

c) __Al + __HCl → __$AlCl_3$ + __H_2

d) __ZnS + __O_2 → __ZnO + __SO_2

e) __NH_4Cl + __$Ca(OH)_2$ → __$CaCl_2$ + __NH_3 + __H_2O

f) __C_4H_{10} + __O_2 → __CO_2 + __H_2O

g) __H_3PO_4 + __NaOH → __Na_3PO_4 + __H_2O

h) __C_2H_2 + __O_2 → __CO_2 + __H_2O

i) __$KClO_3$ → __KCl + __O_2

j) __C_2H_6 + __O_2 → __CO_2 + __H_2O

k) __N_2 + __H_2 → __NH_3

l) __N_2H_4 + __O_2 → __N_2 + __H_2O

m) __Na + __Cl_2 → __NaCl

n) __Fe + __O_2 → __Fe_2O_3

1.6 Learning Check

Mole Relationships in a Chemical Reaction

A chemical reaction shows you the relative number of moles of all species in a chemical reaction. It's like a recipe. Two parts hydrogen plus one part oxygen makes two parts water. – It's just that our "parts" are moles.

$$2H_2(g) + O_2(g) \rightarrow 2H_2O$$

$$\frac{H_2}{O_2} = \frac{2}{1}$$

The coefficients in the balanced reaction give us the mole ratio of any two species, so, if you know the number H_2, you can solve for O_2.

Example

How many moles of O_2 are required to react with 6.284 mol of H_2?

$$\frac{H_2}{O_2} = \frac{2}{1} = \frac{6.284 \text{ mol}}{x}$$

$$2x = 6.284 \text{ mol}$$

$$x = 3.142 \text{ mol of } O_2$$

It doesn't matter which two species you are concerned with, you always know one of them and the ratio will give you other one.

Example

How many moles of each product are produced from 32 mol of nitric acid?

$$3Cu(s) + 8HNO_3(aq) \rightarrow 3Cu(NO_3)_2(aq) + 2NO(g) + 4H_2O(l)$$

$\frac{HNO_3}{Cu(NO_3)_2} = \frac{8}{3} = \frac{32}{x}$	$\frac{NO}{HNO_3} = \frac{2}{8} = \frac{x}{32}$	$\frac{HNO_3}{H_2O} = \frac{8}{4} = \frac{32}{x}$
$8x = 3(32 \text{mol})$ of $Cu(NO_3)_2$	$8x = 2 \times 32 \text{mol}$ of NO	$8x = 4 \times 32 \text{mol}$ of H_2O
$x = 12 \text{ mol of } Cu(NO_3)_2$	$x = 8 \text{ mol of } NO$	$x = 16 \text{ mol of } H_2O$

Note: Notice that it doesn't matter which is on the top of the fraction – you are going to cross multiply anyway. You do need to keep things lined up though! - In the example, HNO_3 is always lined up with 8 and 32.

1.7 Learning Check

1. Ammonia, NH_3, in produced by the Haber Process as follows

$$N_2(g) + 3H_2(g) \rightleftharpoons 2NH_3(g)$$

How many moles of the following are required to make 5.0 mol of ammonia?
a) nitrogen   b) hydrogen

2. Propane, C_3H_8 burns in oxygen according to…

$$C_3H_8 + 5O_2 \rightarrow 3CO_2 + 4H_2O$$

a) How many moles of oxygen are required to react with…
   i) 3.0 mol of C_3H_8?   ii) 20.0 mol of C_3H_8?   iii) 0.50 mol of C_3H_8?
b) If 50.0 mol of O_2 are available, what is the maximum amount of propane that can be burned?
c) How many moles of water are produced in each question in a)?

3. The combustion of octane, a major component in petrol is represented by the following…

$$2C_8H_{18} + 25O_2(g) \rightarrow 16CO_2(g) + 18H_2O(g)$$

a) How many moles of oxygen are required to react with 5.00 mol of octane?
b) How many moles of octane must be burned to produce 7.2 mol of water?
c) i) If 2.20 mol of carbon dioxide were produced from the reaction, how many moles of water are produced at the same time?
   ii) How many moles of octane must have been burned?

Mass Relationships in a Chemical Reaction

Simple stoichiometry questions always follow the same three steps, and you have done them already, you just combine a mass - mole conversion and then apply the ratio from the balanced equation. The molar mass is always for ONE mole. ***Do not apply the coefficient to the molar mass.*** That would mean that the molar mass of hydrogen depends on what it was reacting with - it doesn't. The ratio of coefficients is used to convert from moles of one species into moles of another species.

Common Mistake

Consider the reaction of 24 g of hydrogen in each of the following reactions:

$$2H_2(g) + O_2(g) \rightarrow 2H_2O(g)$$

$$3H_2(g) + N_2(g) \rightarrow 2NH_3(g)$$

WRONG – It's wrong to say that you have 24 g ÷ 4 g•mol^{-1} = 6 mol in the water reaction and 24 g ÷ 6 g•mol^{-1} = 4 mol of hydrogen in the ammonia reaction. The moles of hydrogen you have are independent of the reaction it is undergoing.

CORRECT - Because you have the same number of grams in both, you have the same number of moles in both – i.e. 24 g ÷ 2 g•mol^{-1} = 12 moles of hydrogen.

The ratio of the moles of hydrogen to oxygen or nitrogen is 2:1 or 3:1 respectively and is used to find out information about the other species in the reaction, in this case, nitrogen or oxygen or the products.

> 1. Divide by the molar mass of what you have in grams (to get moles)
> 2. Convert moles of X to moles of Y by applying the ratio from the balanced equation.
> 3. Multiply by the molar mass of Y to obtain the mass of Y.
>
>   **It's simple: Divide, Ratio, Multiply.**

1. Ethanol can be used as a supplement in petrol to make "gasohol". Ethanol burns according to ...

    $$C_2H_5OH + 3O_2 \rightarrow 2CO_2(g) + 3H_2O(g)$$

    a) What mass of oxygen is required to react with 1200 g of ethanol?
    b) If 655 g of water is produced, what mass of ethanol was burned?

1.8 Learning Check

2. Iron(III) oxide may be obtained from iron(II) sulfide by "roasting" the ore in oxygen.

    $$4FeS(s) + 7O_2(g) \rightarrow 2Fe_2O_3(s) + 4SO_2(g)$$

    What mass of iron (III) oxide can be obtained by the roasting of 774g of the sulphide?

3. Silver is used for jewellery and tableware. It becomes tarnished when exposed to small amounts of H_2S.

    $$4Ag(s) + 2H_2S(g) + O_2(g) \rightarrow 2Ag_2S(s) + 2H_2O(g)$$

    What mass of silver sulphide would form from the reaction of 0.015 g of silver?

4. The Ostwald process is how nitric acid is formed. The first step is the reaction of ammonia and oxygen in the presence of a catalyst. The equation is...

    $$4NH_3(g) + 5O_2(g) \xrightarrow{catalyst} 4NO_2(g) + 6H_2O(g)$$

    a) What mass of oxygen is required to completely react with 1.22 kg of ammonia?
    b) What mass of $NO_2(g)$ is produced at the same time?

5. The metal tungsten, which is used in light bulbs can be obtained by heating its oxide with hydrogen

    $$WO_3(s) + 3H_2(g) \rightarrow W(s) + 3H_2O(g)$$

    a) What mass of tungsten can be obtained from 250.0 g of tungsten(VI) oxide?
    b) What mass of hydrogen is required for part a)?

Limiting Reactant

The dreaded question, but also a popular one with IB Examiners because they can get more concepts covered for less candidate time.

As far as you are concerned, it's the same as the previous question, but now you have been given two (or more) reactants and you need to decide which one (there is always **only** one) will give you the correct amount of product.

1. Determine the number of moles of **each** reactant
2. Determine the limiting reactant – this is the new step
3. Use the limiting reactant to find the moles of the product (just like before)
4. Calculate the mass of the product. – just like before.

So how do we do the new step?

There are lots of ways, but remember the limiting reactant is determined to be the reactant which has the fewest number of moles *in terms of the molar ratio of reactants.* (It's the last bit that lots of students forget)

The easiest way to compare the number of moles of each reactant with respect to the molar ratio required is to do a quick check by forcing the ratio to be against 1. In order to do this, divide the moles of each reactant by its coefficient in the balanced equation.

THIS IS A CHECK ONLY – YOU DO NOT USE THIS NUMBER FOR ANY CALCULATIONS.

Example

What mass of aluminium chloride can be produced by the reaction of 50.00 g of Al and 175.0 g of Cl_2?

$$2Al(s) + 3Cl_2(g) \rightarrow 2AlCl_3(s)$$

Solution

$$n_{Al} = \frac{50.00}{26.98 \text{ g mol}^{-1}} = 1.853224611 = 1.853 \text{ mol } (4 \text{ s.f.})$$

$$n_{Cl_2} = \frac{175.0}{70.90 \text{ g mol}^{-1}} = 2.468265162 = 2.468 \text{ mol } (4 \text{ s.f.})$$

The New Step

Now we know the moles of each reactant, but we need to find the lesser with respect to the ratio in the balanced equation so we do the following quick check.

$$\frac{1.853 \text{ mol of Al}}{2} = 0.9265 \qquad \frac{2.468 \text{ mol of } Cl_2}{3} = 0.8227$$

Because the value for Cl_2 is the lesser, this is our limiting reactant, and Al is "in excess" (INXS). Alternately, we can see that we require three times as much chlorine as aluminium - do we have it? - NO, the amount of chlorine is less than what we need, therefore it is the limiting reactant. Now we use the limiting reactant for everything, and all other amounts are relative to it.

Continue as before

To answer the question finally...

$$\frac{product}{limiting \ reactant} = \frac{AlCl_3}{Cl_2} = \frac{2}{3} = \frac{x}{2.468 \text{ mol}}$$

$$3x = 2 \times 2.468 \text{ mol}$$

$$x = 1.646 \text{ mol of } AlCl_3$$

It doesn't matter which way you have the fraction as long as all species are lined up.

Now, finish the problem by finding the mass of 1.646 moles of aluminium chloride.

$mass = number\ of\ moles \times molar\ mass$

$m = n \times M_r$

$m = 1.646\ mol \times 133.33\ g\ mol^{-1}$

$m = 219.5\ g\ of\ AlCl_3$ (4 s.f.)

---

1. Ethane burns in oxygen as follows

$$2C_2H_6(g) + 7O_2(g) \rightarrow 4CO_2(g) + 6H_2O(g)$$

   If 10.2 g of ethane and 44.6 g of oxygen are mixed and ignited...
   a) determine the moles of ethane and oxygen.
   b) determine the limiting reactant.
   c) determine the mass of carbon dioxide produced.

1.9 Learning Check

2. Sodium phosphate can be prepared by the following...

$$3NaOH + H_3PO_4 \rightarrow Na_3PO_4 + 3H_2O$$

   If 36.0 g of NaOH is reacted with 12.0 g of H_3PO_4...
   a) Determine the limiting reactant.
   b) What mass of sodium phosphate should be produced?

3. 100.0 g of each oxygen and butane(C_4H_{10}), are combusted.
   a) Write the balanced equation for the combustion reaction.
   b) Identify the amount of each reactant in moles.
   c) Identify the limiting reactant.
   d) Identify the masses of both products formed.

Determination of Formulae - Gravimetric Analysis

By measuring the changes in mass(gravimetric analysis) in a chemical reaction we can deduce the changes in the mole ratio of atoms. There are two classic examples.

Determine the formula of an oxide of an element. By knowing the mass of a pure element, burning it in oxygen, and then determining the mass of the product we can work out how much oxygen was incorporated into the compound.

6.98 g of iron was heated in oxygen and the resulting product had a mass of 9.98 g. Determine the formula of the iron oxide formed.

	iron	oxygen
mass	6.98 g	9.98-6.98 = 3.00 g
moles	6.98 g / 55.85 g mol-1 0.125 mol	3.00 g / 16.00 g mol-1 0.1875
ratio - divide by smaller	0.125/0.125 = 1	0.1875/0.125 = 1.5
clear fraction	2	3

Therefore the formula of the oxide is Fe_2O_3.

The second example is the determination of the number of water molecules present in the lattice of a hydrated salt eg $CuSO_4 \cdot 5H_2O$ contains five water molecules for each copper atom. Upon strong heating the lattice is broken down, and water is released from the hydrate, forming the anhydrous salt.

Example

A 6.16 g sample of hydrated magnesium sulfate was heated until no further mass change occurred and the anhydrous product had a mass of 3.01 g. Determine the formula of the hydrate.

	anhydrous salt, $MgSO_4$	water
mass	3.01 g	6.16 g - 3.01 g = 3.15 g
moles	3,01 g / 120.38 g mol^{-1} = 0.0250 mol of $MgSO_4$	3.15 g / 18.02 g mol^{-1} = 0.175 mol of H_2O
ratio	0.0250 mol / 0.0250 mol = 1	0.175 mol / 0.025 mol = 7

The formula of the hydrate is $MgSO_4 \cdot 7H_2O$, magnesium sulfate heptahydrate

Mixtures & Solutions

Unlike a chemical compound which has a specific mole ratio (e.g. H_2O), a mixture or solution may have any ratio. A mixture has two or more distinct visible components, such as oil & water. Milk contains small droplets of fats seen under a microscope. This gives milk its opacity.

A solution is a transparent homogeneous mixture - there is only one uniform component, where the solute is distributed throughout the solvent.

A solution contains a solute dissolved in a solvent, and the molar concentration is expressed as moles of solute per cubic decimetre of solution (not solvent).

A solution is made by dissolving the required amount of solute in a minimum of solvent, and then adding what ever amount of solvent is required to reach the desired volume.

Definition

The molar concentration is called molarity. It may be written in two different ways. The fundamental units "mol dm^{-3}" or simply "M" (Which is often underlined or **bold**). E.g. The student was using a 0.25 mol·dm^{-3} solution = 0.25 **M** solution.

$$\text{concentration} = \frac{\text{number of moles}}{\text{volume of solution}}$$

$$C = \frac{n}{V}$$

where C is the concentration in mol dm^{-3}, n is the number of moles, and V is the volume in dm^3.

IB and most other organisations are moving towards using dm^3 as the basic unit of volume. However the term "liter" is still popular in language and texts.

One liter is defined as 1 dm^3.

$$1 \text{ dm}^3 = 1.0 \text{ L} = 1000 \text{ mL} = 1000 \text{ cm}^3$$

Quantitative Chemistry

Making solutions and determining solute mass

You need to be able to state how to make a certain volume and concentration of solution.

Very often a solution of a known concentration - a standard solution - is used to determine the amount of another substance by a known chemical reaction - i.e. titration.

What mass of $NaHCO_3$ is required to make 250 cm³ of a 0.350 M solution?

$n = CV$
$n = 0.350 \text{ mol dm}^{-3} \times 0.250 \text{ dm}^3$
$n = 0.0875$ mol of $NaHCO_3$ required
mass $= n \times M_r$
mass $= 0.0875$ mol $\times 84.01$ g mol⁻¹
mass $= 7.35$ g $NaHCO_3(s)$ required
Dissolve 7.35 g of $NaHCO_3(s)$ is some water and add water up to 250 cm³.

Example

Solution

Determine the concentration of the following solutions.
a) 250 cm³ containing 45.0 g of $MgSO_4$
b) 0.100 dm³ containing 10.0 g of NaCl
c) 500 cm³ containing 2.42 g of $Fe(NO_3)_3$
d) 25.0 cm³ containing 0.210 g of $NaHCO_3$

1.10 Learning Check

Determine the mass of solute required to make the following solutions.
a) 1.00 dm³ of 0.25 M KI
b) 250.0 cm³ of 0.1 M $AgNO_3$
c) 100.0 cm³ of 0.25 M NaOH
d) 0.500 dm³ of 0.40 M $Mg(NO_3)_2$

Kinetic Molecular Theory

The Kinetic Molecular Theory states that ...

1) Particles are in constant random motion and...
2) That motion depends upon the ***absolute*** temperature measured in Kelvin.

Simply put, absolute zero means there is no particle motion – all translational, rotational and vibrational motion stops. So you can't go below absolute zero, because once motion is stopped, it can't be more stopped!!

Temperature is therefore a measure of kinetic energy or motion of particles. Greater temperature means that the particles have greater kinetic energy.

The absolute temperature scale is measured in Kelvin. **Zero Kelvin means zero kinetic energy.**

Pressure

Pressure - the force of the particles' collisions on the container walls.

Definition

Pressure can be measured in many different ways, all of which have some historical basis. However, the two common pressure units in IB Chemistry are kilopascals (kPa) and atmospheres (atm).

$$1 \text{ atm} = 100 \text{ kPa} = 1.0 \times 10^5 \text{ Pa}$$

Boyle's Law

Boyle's Law is the relationship between volume and pressure at constant temperature. Pressure and volume are inversely proportional. As the volume decreases, the particles are forced closer together and so, the number of collisions increases, and thus pressure increases.

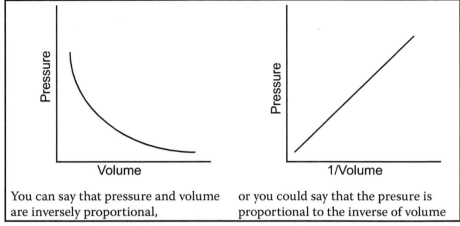

You can say that pressure and volume are inversely proportional, or you could say that the presure is proportional to the inverse of volume

Figure 1.2: Pressure and Volume graphical relationships.

$$\text{Pressure} \propto \frac{1}{\text{Volume}}$$

$$P \propto \frac{1}{V}$$

$$PV = \text{constant}$$

$$(PV)_{\text{condition 1}} = \text{constant} = (PV)_{\text{condition 2}}$$

Boyles' Law Expression

$$\boxed{P_1 V_1 = P_2 V_2}$$

Pressure Law

This is the relationship between pressure and temperature at constant volume.

As temperature increases, the kinetic energy of the particles increases, therefore they are colliding with the sides of the container with greater frequency and force. Both of these factors cause the pressure to increase.

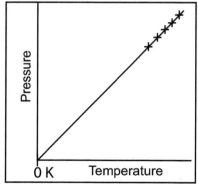

Figure 1.3: Pressure and temperature relationship

If the particles have zero kinetic energy at zero Kelvin, then they can't collide with the container walls.

Charles' Law

Charles' Law is the relationship between volume and temperature at constant pressure. As temperature increases, so does the kinetic energy of the particles. This causes them to collide more often and more violently with each other, and therefore spread out.

$$\text{Volume} \propto \text{Temperature}$$

$$V \propto T$$

$$\frac{V}{T} = \text{constant}$$

$$\left(\frac{V}{T}\right)_{\text{condition 1}} = \text{constant} = \left(\frac{V}{T}\right)_{\text{condition 2}}$$

$$\boxed{\frac{V_1}{T_1} = \frac{V_2}{T_2}}$$

Expression

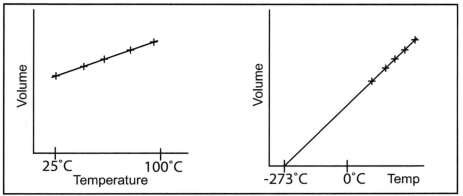

Figure 1.4: Volume and Temperature graphical relationships

Generally, volume is proportional to temperature, however, zero volume can only exist at zero KELVIN, not zero Celsius. - see Ideal Gases.

Watch out for changes involving a doubling or tripling of Celsius degrees, the proportionality only works for the absolute (Kelvin) scale.

Exam Trap

The Combined Gas Law

If we put Boyle's Charles' and the pressure law together we get,

$$\boxed{\frac{P_1 V_1}{T_1} = \frac{P_2 V_2}{T_2}}$$

How will the volume of an ideal gas change when the pressure is quadrupled and the absolute temperature is tripled?

Example

$$\frac{P_1 V_1}{T_1} = \frac{(4P_1) V_2}{(3T_1)}$$

$$\frac{P_1 V_1 (3T_1)}{(4P_1) T_1} = V_2$$

$$\frac{3}{4} V_1 = V_2$$

The new volume, V_2 will be ¾ of the original volume, V_1.

Avogadro's Law

Up until now, we have been concerned only with a fixed amount (or mass) of gas. Now we consider how changing the number of moles (amount) of gas affects its volume. Avogadro's Law is the relationship between number of particles and volume.

> Avogadro's Law: Equal volumes of gas (under equal conditions) must contain an equal number of moles of gas.

Or put another way, volume is proportional to the number of moles at constant pressure, or pressure is proportional to the number of moles at constant volume.

The identity of the gas doesn't matter because the gas particles are so spread out.

Avogadro's Law of Combining Volumes

A popular type of multiple-choice question on Paper 1.

Avogadro's Law simply means that you can treat moles and volume or moles and pressure as proportional if all other factors remain constant.

Classic Question

Consider the Haber Process for the production of ammonia.

$$N_2(g) + 3H_2(g) \rightleftharpoons 2NH_3(g)$$

What volume of ammonia will be produced by the reaction of 4 dm³ of N_2 and 9 dm³ of H_2?

Consider this as a limiting reactant question. The molar ratio of N_2 to H_2 is 1:3. Is 9 dm³ three times greater than 4 dm³? - NO! We need 12 moles of H_2. We only have 9, so H_2 is the limiting reactant, and we always use the limiting reactant to figure out the amount of product.

$$\frac{\cancel{H_2}}{N\cancel{H_3}} = \frac{3}{2} = \frac{9}{x}$$

$$x = 6$$

$$NH_3 = 6\ dm^3$$

Standard Molar Volume

Standard molar volume is the volume of one mole of any gas at standard temperature and pressure for a gas - 1 atm and 273 K (0°C).

> 1.0 mol of any gas occupies 22.4 dm³ @ Standard Conditions (STP)

Do not confuse with the value for RTP - "room" temperature and pressure, which is 25°C, not 0°C, so the volume is bigger - 24.0 dm³.

Also, don't confuse with "Standard Conditions" for Energetics - which use 25°C for "Standard Temperature - They're not that standard after all.

Whenever you have been given information about gases, check to see if they are at Standard conditions (1 atm & 273K) - If it is, then you don't need to use PV=nRT, you can simply the standard molar volume of V=22.4 dm³mol⁻¹!

Example

A sample of 0.0412 g of a gas has a volume of 35.4 cm³ at 1 atm and 273K, what is the molar mass of the gas?

Ideal Gas Law

The Ideal gas law combines all of the fore-mentioned relationships.

An ideal gas is a gas that does **not** have any...

1. molecular volume nor any
   - so that the volume of a gas can go to zero
   - (ignore volume of particles)

2. inter-particle forces
   - so that the gas never condenses to a liquid
   - (ignore IMFs)

Most gases behave as Ideal Gases because the volume of the gas is usually so much bigger than the volume of the actual molecules that it is negligible, and they are at temperatures far above the boiling/condensation points that the IMFs are insignificant.

Gases exhibit "non-ideal" behaviour when they are under high pressures, and low temperatures. This can be seen as the gas condenses into a liquid with a measurable volume.

The ideal gas law relationship is:

$$PV = nRT$$

You must make sure that you pay attention to units. — *Exam Trap*

Most students learn that the value for R; the ideal gas constant is 8.314, but many forget to include the units of $kPa \cdot dm^3 \cdot mol^{-1} \cdot K^{-1}$. There are other values of R that relate to other units of pressure and volume.

Becareful because in the databook, the units are given as $J \cdot mol^{-1} \cdot K^{-1}$. The fact is that the product of pressure an volume is Joules, and can be expressed as $kPa \cdot dm^3$ or $Pa \cdot m^3$. Watch out that the units are lined up correctly.

The most common mistake is to not pay attention to the units in the question. Lots of times, exams give information in similar units like Pa not **kPa**, and if you don't convert your answer is off by a factor of 1000, which can lead to some strange answers. — *Common Mistake*

Calculate the molar mass of a volatile liquid / gas. — *Classic Question*

You are given the mass of a gas, the volume and conditions, and must find the molar mass. As molar mass is only the mass per mole, you have all the information you need. Mass is given and you can rearrange PV=nRT to find number of moles, n.

An unknown gas of mass 0.625 g occupies a volume of 351 cm³ at 25°C and 1x10⁵ Pa. Calculate the molar mass of the gas. — *Example*

Don't forget to convert quantities to the correct units!!!

25°C = 298K
1.00x10⁵ Pa = 100 kPa
351cm³ = 0.351 dm³

$$PV = nRT$$

$$n = \frac{PV}{RT}$$

$$n = \frac{(100\ kPa)(0.351\ dm^3)}{(8.314\ kPa \cdot dm^3 \cdot mol^{-1} \cdot K^{-1}) \times (298\ K)}$$

$$n = 0.0142\ mol$$

$$molar\ mass = \frac{mass}{moles} = \frac{0.625\ g}{0.0142\ mol} = 44.12\ g \cdot mol^{-1}$$

Remember to use common sense - the molar mass of any volatile gas is likely going to be between 20 and 200 $g \cdot mol^{-1}$.

Yield: Theoretical, Experimental and Percentage

The theoretical yield is simply the answer to a stoichiometry question – it's how much product you are supposed to get at the end of your experiment

The experimental yield is what you actually do get. In most cases the experimental yield is less than the theoretical yield, but not always.

Reasons for low yield	Reasons for high yield
loss during transfer	insufficient drying
equilibrium / reaction did not go to completion	gain of oxygen (oxidation)
side reactions	side reactions
impure reactants (already reacted somewhat)	

Definition

Percentage yield is the ratio of experimental to theoretical expressed as a percent.

$$\text{Percent Yield} = \frac{\text{Experimental Yield}}{\text{Theoretical Yield}} \times 100\%$$

1.11 Learning Check

1. A student determines that the theoretical yield of her preparation of aspirin should produce 4.3 g of product. After drying and weighing her product, she obtained 3.8 g. What was her percent yield?

2. Sodium thiosulphate may be produced by boiling solid sulphur, $S_8(s)$ in a solution of sodium nitrite, $Na_2SO_3(aq)$, according to the reaction

$$S_8(s) + 8Na_2SO_3(aq) \rightarrow 8Na_2S_2O_3(aq)$$

If a student starts with 15.50 g of sulphur and an excess of Na_2SO_3, determine the theoretical yield of product. If only 62.5 g of product is collected, determine the percent yield.

Empirical & Molecular Formulae

A formula is simply the molar ratio of elements in a compound. Water has 2 moles of hydrogen for every mole of oxygen. This ratio is fixed for a given compound. If you have a compound of 2 moles of hydrogen for 2 moles of oxygen, you don't have water; you have hydrogen peroxide – a compound with properties different from water

As long as we know the ratio of the elements in any units, we can convert to moles and we can determine the formula.

Lots of time these questions show up on Paper 1, and that's a good thing. Because you can't have a calculator, you must be given information in simple multiples of the relative mass.

Example

What is the empirical formula of a compound that contains 46 g of sodium, 64 g of sulphur and 48 g of oxygen?

If you look at the periodic table, you notice that the molar masses are 23, 32, and 16 respectively. Therefore you must have 2 moles of sodium, 2 moles of sulphur and 3 moles of oxygen. $Na_2S_2O_3$!

Example

Which of the following compounds has the greatest empirical mass?
a) C_6H_6 b) $C_{20}H_{40}$ c) C_4H_{10} d) CH_3

Solution: (c) – the empirical formulae are CH, CH_2, C_2H_5 and CH_3 respectively

Quantitative Chemistry — page 23

Calculations from percentage information

Problem solving steps

1. Assume 100 g of compound – therefore %'s become grams
2. Convert grams to moles by dividing by the molar mass
3. Divide by the smallest value to get whole number ratios

Problem solving steps

Compound Q is analysed and found to contain 85.6% carbon and 14.4% hydrogen. Determine the empirical formula of Q.

Process	Carbon	Hydrogen
Information	85.6% carbon	14.4%
Assume 100g	85.6 g	14.4 g
Divide by M_r	85.6 g ÷ 12.01 = 7.13 mol	14.4 g ÷ 1.01 = 14.26 mol
divide by smallest	7.13 ÷ 7.13 = 1	14.26 ÷ 7.13 = 1.999
ratio	1	2

The formula of compound Q is therefore CH_2.

Be Careful

Sometimes formulae do not have a ratio against one, for example, Fe_2O_3. If you do the previous calculations, and you find that at the end, you have a formula of 1:1.5, then you will have to double all subscripts to attain a whole number ratio of 2:3. Try the following.

1.12 Learning Check

1. Compound X is analysed and found to contain 82.63% carbon and 17.37% hydrogen. Determine the empirical formula of X.

2. If compound Y contains 89.92% carbon and 10.08% hydrogen. What is the empirical formula of compound Y?

3. What is the empirical formula of a compound containing 92.24% carbon and 7.76% hydrogen?

C.Lumsden — IB HL Chemistry

Calculations from Empirical Data

Instead of being given the ratio of elements in terms of a percentage, we may be given masses of combustion products. We can convert to moles from the molar masses of the combustion products. Because carbon dioxide contains one mole of carbon for every mole of CO_2, that will be easy. We must remember that water contains 2 moles of hydrogen for every mole of water, so we must double the number moles of water to determine the number of moles of hydrogen.

> 1. Determine the number of moles of carbon from carbon dioxide.
> 2. Determine the number of moles of hydrogen from water (multiply by 2).
> 3. Force the ratio against 1 by dividing by the smallest number of moles.
> 4. Clear any fractions if necessary.

Example

A sample of a hydrocarbon with a mass of 2.37 g was burned in excess O_2 to produce 7.18 g of CO_2 and 3.67 g of H_2O. What is the empirical formula of the hydrocarbon?

Thought Process	Carbon solution	Hydrogen solution
Determine moles of combustion products: CO_2 & H_2O	$n_{CO_2} = \dfrac{7.18\ g}{44.01\ gmol^{-1}}$ $n_{CO_2} = 0.163\ mol$	$n_{H_2O} = \dfrac{3.67\ g}{18.02\ gmol^{-1}}$ $n_{H_2O} = 0.204\ mol$
moles of elements (double for hydrogen)	$n_C = 0.163\ mol$	$n_H = 2 \times 0.204\ mol$ $n_H = 0.408\ mol$
divide by smallest	$0.163\ mol \div 0.163 = 1$	$0.408\ mol \div 0.163 = 2.50$
clear fraction (x2)	2	5

The empirical formula is therefore C_2H_5!

1.13 Learning Check

1. 2.75 g of a compound containing only carbon and hydrogen were combusted. The combustion products were 8.05 g of CO_2 and 4.94 g of H_2O. Determine the empirical formula of the compound.

2. 0.875 g of a hydrocarbon were burned in excess oxygen and produced 2.74 g of CO_2 and 1.12 g of water. What is the formula of the hydrocarbon?

3. A hydrocarbon sample of mass 1.15 g is completely combusted to form 3.44 g of carbon dioxide and 1.88 g of water. Determine the empirical formula of the sample.

Compounds containing oxygen

If you have been given percentage information, then you can work as before but now you have three elements to deal with.

However if you are given empirical data, it's not so easy. The problem is that you have to determine the number of moles of oxygen in the compound, but that the presence of oxygen gas for combustion is confounding because the oxygen present in the two products has come from two sources.

$$C_xH_yO_z + O_2(g) \rightarrow CO_2(g) + H_2O(g)$$

The solution, therefore, is to ignore the amount of oxygen in the products and calculate the amount of oxygen in the fuel (reactant) by subtracting what we do know – the carbon and the hydrogen – from the starting mass.

Let's use the combustion of ethanol, C_2H_6O, as a known example.

Example

A sample of 2.35 g of compound containing C, H and O, is combusted in an excess of oxygen to produce 4.49 g of CO_2 and 2.75 g of H_2O. Determine the formula of the compound.

Thought Process	carbon solution	hydrogen solution	oxygen solution
Determine moles of combustion products CO_2 & H_2O	$n_{CO_2} = \dfrac{4.49\ g}{44.01\ gmol^{-1}}$ $n_{CO_2} = 0.102\ mol$	$n_{H_2O} = \dfrac{2.75\ g}{18.02\ gmol^{-1}}$ $n_{H_2O} = 0.153\ mol$	
moles of elements (double for H)	$n_C = 0.102\ mol$	$n_H = 2 \times 0.153\ mol$ $n_H = 0.305\ mol$	
Mass of element from compound Subtract for oxygen.	$m_C = 0.102 \times 12.01$ $m_C = 1.23\ g$	$m_H = 0.305 \times 1.01$ $m_H = 0.308\ g$	$m_O = 2.35\ g$ $-(1.23g + 0.308g)$ $m_O = 0.812$
moles of elements	as above $n_C = 0.102\ mol$	as above $n_H = 0.305\ mol$	$n_O = 0.812 \div 16.00$ $n_O = 0.0508\ mol$
divide by smallest	$0.102\ mol \div 0.0508$ $= 2$	$0.305\ mol \div 0.0508$ $= 6$	$0.0508\ mol \div 0.0508$ $= 1$
mole ratio	2	6	1

Therefore the formula (as expected) is C_2H_6O.

1.14 Learning Check

1. 4.35 g of a compound containing carbon, oxygen and hydrogen were combusted and the products were 6.38 g of carbon dioxide and 2.61 g of water. What was the empirical formula of the compound?

2. An alcohol of mass 2.63 g was completely combusted in an excess of oxygen. The products were found to be 5.78 g of CO_2 and 3.15 g of H_2O. Determine the empirical formula of the compound.

3. One gram of a carboxylic acid is analysed by combustion and the reaction produced 2.00 g of CO_2 and 0.818 g of H_2O. What is the empirical formula of the acid?

Molecular Formula

The empirical formula represents the simplest ratio of elements. Usually molecules are more complex.

The molecular formula is an integer multiple of the empirical formula. Below are two examples for CH_2 and CH_2O empirical formulae.

Empirical Formula	CH_2	CH_2O	1
Possible Molecular Formulae	C_2H_4	$C_2H_4O_2$	2x
	C_3H_6	$C_3H_6O_3$	3x
	C_4H_8	$C_4H_8O_4$	4x
	C_5H_{10}	$C_5H_{10}O_5$	5x
	C_6H_{12}	$C_6H_{12}O_6$	6x

Table 1.5: Relationship between empirical and molecular formulae

Be Careful! The empirical formula is different from the general formula for organic compounds.

You are often asked to determine the molecular formula once you have determined the empirical formula (or the information is given). You only need one more piece of information - the molar (or molecular) mass.

The key is to find the mass of the empirical unit. The molar mass must be a whole number multiple of the empirical mass.

Empirical Mass	$CH_2 = 14$	$CH_2O = 30$	1
Possible Molecular Formulae masses	$C_2H_4 = 28$ (2 x 14)	$C_2H_4O_2 = 60$ (2 x 30)	2
	$C_3H_6 = 42$ (3 x 14)	$C_3H_6O_3 = 90$ (3 x 30)	3
	$C_4H_8 = 56$ (4 x 14)	$C_4H_8O_4 = 120$ (4 x 30)	4
	$C_5H_{10} = 60$ (5 x 14)	$C_5H_{10}O_5 = 150$ (5 x 30)	5
	$C_6H_{12} = 74$ (6 x 14)	$C_6H_{12}O_6 = 180$ (6 x 30)	6

Table 1.6: Empirical Mass Relationships

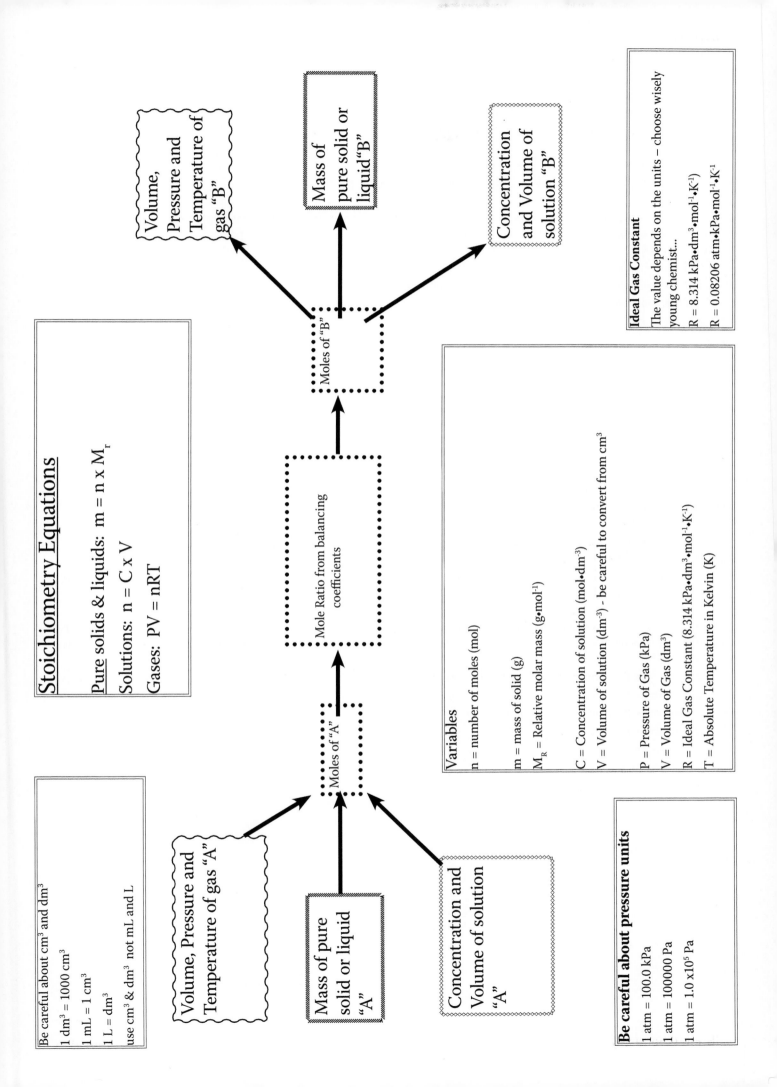

Summary Questions

1. 250 cm³ of a solution of unknown concentration of sodium chloride. A 20.00 cm³ sample is reacted with excess silver nitrate causing a white precipitate. When washed and dried, the precipitate had a mass of 0.430 g.
    a) Write the balanced equation including state symbols.
    b) Determine the moles of precipitate, and hence the moles of sodium chloride in 20 cm³.
    c) Determine the concentration of the solution.
    d) What mass of sodium chloride was used to make the original solution?

2. Calcium carbide, CaC_2(s) reacts with water to produce a flammable hydrocarbon gas, and calcium hydroxide. 0.0405 g of the gas has a volume of 38.65 cm³ at 22.00°C and 98.00 kPa. Upon combustion the same amount of gas produces 0.137 g of CO_2 and 0.0280 g of H_2O.
    a) Determine the empirical formula of the gas.
    b) Determine the number of moles and hence the molar mass of the gas.
    c) Determine the molecular formula.
    d) Write a balanced equation for the reaction of calcium carbide with water.

3. Aspririn is made by the reaction of salicylic acid, $C_7H_6O_3$ and ethanoic anhydride, $C_4H_6O_3$ according to the reaction

$$C_7H_6O_3 + C_4H_6O_3 \rightarrow C_9H_8O_4 + CH_3COOH$$

If 6.00 g of salicylic acid and 5.00 g of ethanoic anhydride are reacted.
    a) Determine the moles of each reactant.
    b) Determine the limiting reactant.
    c) Determine the mass of product that should be produced.
    d) If a student only recovers 6.50 g of aspirin, calculate the theoretical yield.

4. Hexane, C_6H_{14}, is a combustible hydrocarbon. Consider the combustion of a mixture of 4.50 g of hexane and 11.25 g of oxygen.
    a) Write the balanced equation for the reaction
    b) Determine the moles of each reactant.
    c) Determine the limiting reactant.
    d) Determine the mass of each, carbon dioxide and water produced.
    e) Determine the mass of excess reactant.

5. A metal sulphate has the formula M_2SO_4. 10.99 g of the compound was dissolved in water to make 500 cm³ of solution. A 25.00 cm³ sample was removed and reacted with an excess of $BaCl_2$(aq) to produce a precipitate of $BaSO_4$(s), which, when dried had a mass of 1.167 g.
    a) Determine the number of moles of $BaSO_4$(s) precipitated.
    b) Determine the concentration of the M_2SO_4 solution.
    c) Determine the number of moles of M_2SO_4 in the original solution.
    d) Determine the molar mass of M_2SO_4.
    e) Determine the identity of M.

Chapter 2

Atomic Structure

In this chapter...

30	Subatomic Particles
30	The Nuclear Atom & Isotopes
31	Isotope Notation
31	The Mass Spectrometer
32	Relative Atomic Mass
32	Calculating Relative Average Atomic Mass
32	Calculating Natural Abundance
33	The Electronic Atom - Basic structure
33	Evidence for Shells – the Hydrogen spectrum
34	Explaining the Hydrogen Spectrum
35	The Electronic Atom - Shells & Sub-Shells
37	AHL - Energy of Light
38	Orbital Diagrams & Hund's Rule
38	Exceptions to the filling pattern in the d-block.
39	Successive Ionization Energies – evidence for shells
40	Exceptions to Ionization Energy Trend - Evidence for subshells
41	Summary Questions

Subatomic Particles

Atoms are made up of 3 subatomic particles – know their properties...

Particle	Relative Mass	Relative Charge	Location	Actual Mass /kg	Actual Charge /C
Proton	1	+1	Nucleus	1.673×10^{-27}	$+1.602 \times 10^{-19}$
Neutron	1	0 (neutral)	Nucleus	1.675×10^{-27}	0
Electron	5×10^{-4}	-1	Orbiting nucleus	9.109×10^{-31}	-1.602×10^{-19}

Figure 2.1: Subatomic Particles

The protons and neutrons give the nucleus and hence the atoms their mass, while the electron orbits give the atoms their volume. The number of protons is unique for a given element, or if you like, the way we identify elements is by the number of protons. The number of neutrons can be different, which gives rise to different isotopes. Neutrons only help to stabilize the nucleus. An excess or deficit of neutrons leads to radioactive isotopes. The number of electrons can change during the course of a chemical reaction as atoms gain or lose electrons to form different ions or oxidation states.

When working with isotopes, we tend to use the relative masses, where protons and neutrons are the same. In reality a neutron is slightly heavier. In either case, the electrons are about 2000 times less massive than a proton or neutron - when dealing with isotopes, we can consider their mass as negligible, but they really do have a mass

The Nuclear Atom & Isotopes

The nuclear atom was determined by Ernst Rutherford and his team, Geiger and Marsden.

The genius of Rutherford's experiment was that he was able to mathematically eliminate all other possibilities so the only remaining model was one in which the atom had a nucleus which concentrated all of the mass and positive charge of the atom into a volume 10000 times smaller than the total volume of the atom.

Isotopes were proposed and discovered within the same year, 1913. The first evidence of isotopes was of radioactive elements which did not occupy a new place in the periodic table. As they were "in the same place" as other elements they were given the term "isotope".

J.J. Thomson's experiments in the same year were the first pieces of evidence for non-radioactive isotopes. Six years later, in 1919 Aston discovered that the isotopes were always of whole number mass by using a mass spectrometer.

Isotopes are... (here are two ways to say the same thing...)

atoms with the same number of protons with a different number of neutrons.
atoms of the same element with different masses.

Get the point! you must say that isotopes are ***atoms***.

Thinking: There isn't one or "real" or "true" atom and the others are isotopes. Consider twins - there is no "real" person and the sibling is a copy. They are twins of each other. There may be one isotope that is more ***abundant*** – there is a greater percentage of that one compared to others.

Most elements have multiple naturally occurring isotopes, but not all. Fluorine, for instance has only one naturally occurring isotope – Fluorine-19. Whereas, chlorine has two, chlorine-35 and chlorine-37

As a teacher I always like to ask questions about isotopes that have a different mass from the average mass found on the periodic table – I always catch at least one student saying: "I think this question is wrong." Go back to the definition.

Isotopes differ from each other due to the change in mass (only). Isotopes may have slightly different physical properties. The isotope of hydrogen containing one neutron is called "deuterium".

Atomic Structure

When water is made from deuterium instead of normal hydrogen, it has a higher boiling point of 101.4°C. The rate of diffusion is another property that changes because the heavier isotopes will move more slowly. This is the basis of how isotopes are separated to gnerate an isotopically pure sample, and is key in isolating the useful isotope of uranium from the others. Isotopes have the same chemical properties because the chemical properties are determined by the nuclear charge and electronic structure, not the mass.

Isotope Notation

$^{A}_{Z}X$ Where Z is the number of protons(atomic number) and A is the number of protons plus neutrons(atomic mass). At the moment, we will deal with neutral atoms, so electrons=protons.

Complete the following table.

Isotope symbol	Atomic Mass (A)	Atomic Number (Z)	Protons	Neutrons	Electrons
$^{23}_{11}Na$	23	11	11	12	11
$^{1}_{1}H$					
$^{2}_{1}H$					
$^{3}_{1}H$					
$^{10}_{5}B$					
$^{11}_{5}B$					
$^{35}_{17}Cl$					
$^{37}_{17}Cl$					

2.1 Learning Check

The Mass Spectrometer

The mass spectrometer is a device that separates particles based on their mass.

The operation of the mass spectrometer is not part of the core syllabus, however for now, we can consider it as simply a machine that counts how many of each type of isotope there is in a sample. The results of a mass spectrometer may be presented graphically as a "spectrogram" where you can identify the releative amounts of ions by the heights of the peaks. We are interested in the ratio of each isotope to each other, called the *natural abundance*.

The mass spectrometer measures the "mass to charge" ratio of the atoms. For our purposes this can be regarded as mass for the moment, as we can assume the charge will be +1.

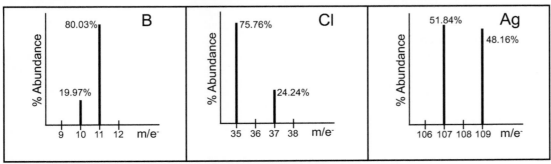

Figure 2.2: Mass Spectrum examples

Relative Atomic Mass

Carbon-12, which has 6 protons, 6 neutrons and 6 electrons is defined as having a relative atomic mass of exactly 12.000. One "unified atomic mass unit" (amu or u) is defined as 1/12 of this mass. All other isotopes are measured compared to this value.

The relative mass is the ratio of the average mass of the atom to the unified atomic mass unit, which is one twelfth of the mass of a carbon-12.

> The relative atomic mass is the ratio of the average mass of the atom to the unified atomic mass unit, which is one twelfth of the mass of a carbon-12.

If two isotopes are present in equal amounts, then the average atomic mass is going to be the simple average. If one isotope is present in a greater proportion, it counts more to the average, so we use the weighted average mass. The average relative atomic mass is what you read off of the periodic table.

Calculating Relative Average Atomic Mass

You need to multiply each atomic mass by its relative abundance and add the numbers up!

Example

Chlorine exists as two naturally occurring isotopes; Cl-35 and Cl-37. If the Cl-35 has an abundance of 75.76% and the remainder is Cl-37, determine the relative atomic mass of chlorine.

Solution

$$A_r = \frac{\text{Mass}_{Isotope(1)} \times \%\text{abundance} + \text{Mass}_{Isotope(2)} \times \%\text{abundance}}{100}$$

$$A_r = \frac{(35 \times 75.76\%) + (37 \times 24.24\%)}{100\%}$$

$$A_r = 35.48$$

2.2 Learning Check

1. Silver has 2 naturally occurring stable isotopes, in the following ratios: ^{107}Ag: 51.84% and ^{109}Ag: 48.16%. Determine the relative atomic mass of silver.

2. Magnesium has three naturally occurring isotopes, in the following abundances: ^{24}Mg: 79.0%, ^{25}Mg: 10.0%, ^{26}Mg: 11.0%. Calculate the relative atomic mass of magnesium.

3. Boron exists as 19.97% ^{10}B and 80.03% ^{11}B. Determine the A_r.

Calculating Natural Abundance

Calculating natural abundance isn't really a chemistry question, it's a math question. Let's take a common example - chlorine.

Example

Chlorine exists as two isotopes - ^{35}Cl and ^{37}Cl. If the average atomic mass is 35.45, calculate the percentage calculation for each isotope.

Solution

Let the fraction of Cl-37 be x
Then the fraction of Cl-35 will be (1-x)
37x + 35(1-x) = 35.5
37x + 35 - 35x = 35.5
2x = 0.5
x = 0.25
Therefore ^{37}Cl has 25% abundance, and ^{35}Cl has the rest - 75%

Atomic Structure

2.3 Learning Check

1. Thallium consists of thallium-203 and thallium-205. Using the value from the periodic table, determine the relative abundance of each isotope.

2. Lithium consists of ^{6}Li and ^{7}Li. Calculate the percent ratio of the isotopes.

3. Gallium is made up of ^{69}Ga and ^{71}Ga. Determine the isotopic ratio using the average value from the periodic table.

The Electronic Atom - Basic structure

Now that you have studied the nucleus with the two "nucleons" – protons and neutrons, we look at the structure of the atom in terms of electrons.

Electrons are arranged in shells or orbits or layers, and are kept in their orbits by their negative charge and the attractive force of the positive nucleus.

There are three pieces of evidence that lead us to understand the arrangement of the electrons around the nucleus - the electronic configuration.

Evidence for Shells – the Hydrogen spectrum

A continuous spectrum contains **all** the wavelengths / frequencies / colour or **ENERGIES** of electromagnetic radiation. (Not just light.) The visible part of the continous spectrum is seen when white light is passed through a prism (or a raindrop!)

Figure 2.3: A Continuous Spectrum

A line spectrum contains **only certain** / specific / discrete wavelengths / frequencies / colours or **ENERGIES** of electromagnetic radiation.

When we observe hydrogen in a discharge tube, we see a line spectrum, not a continuous spectrum.

Figure 2.4: The Hydrogen Emission Line Spectrum

Explaining the Hydrogen Spectrum

In hydrogen atoms the electron is in its ground state when the single electron is as close to the nucleus as possible due to the attraction between the positive nucleus and the negative electron.

The electron can be given extra energy by a fast moving electron in the discharge tube or by the energy of a flame, and this causes it to be promoted to a higher energy level. We now call this atom "*excited*".

Due to attraction to the nucleus, the electron "falls" back down. When it does this, it goes to a lower potential energy state and the energy is emitted as electromagnetic radiation (light).

As only certain energies of light are seen, the energy difference between any two orbitals must be a fixed amount. The difference in energy levels (shells) is the same amount of energy as the energies seen in the line spectrum. Therefore the energy of the light emitted directly reflects the spacing between the orbitals.

Stage 1 - Ground State	Stage 2 - Excited State	Stage 3 - Ground State
Orbiting electron is in the ground state, nearest to the nucleus, due to attraction	Electron is knocked up to a higher energy state by collision of a fast moving electron	Electron returns to the ground state and releases extra energy in the form of light.

Figure 2.5: Electronic Transitions in Hydrogen

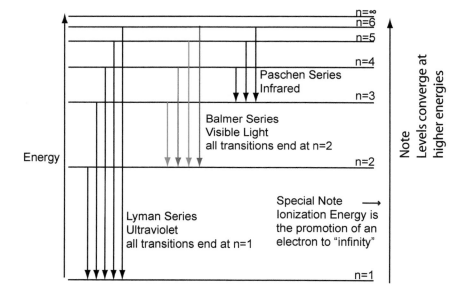

Figure 2.6: Energy Level Transitions

Get the point!

If you look carefully at both the hydrogen emission spectrum and the energy level transitions above, you will notice that as you get towards a higher energy level, the levels converge.

The Electronic Atom - Shells & Sub-Shells

There are some basic rules governing the number and placement of electrons around the nucleus.

Firstly, electrons exist in shells or orbits - you may have learnt this already. The maximum number of electrons in any shell is $2n^2$, where n is the shell number. So the shells can hold 2, 8, 18, 32, 50 and 72 electrons respectively. An orbit or shell is defined by its size.

This can be a lot of electrons to have in one place, so they are organized into pairs of electrons in ORBITALS. Orbitals are regions of space with a certain shape where we are likely to find an electron. Orbitals are arranged in groups called sub-shells depending upon their shape.

A subshell is simply a group of orbitals with the same shape. Some shapes like a sphere have no directon associated with them, while others that are linear or more complex may be oriented along different axes.

The simplest shape of orbital is the "s" orbital which is spherical in shape (the "s" doesn't stand for spherical, but you can remember it that way). The s-orbtial has only one region of space (lobe) where you are likely to find an electron. Because it is spherical, there is no direction associated with it, and therefore there is only on "s" orbital.

The next more complex shape is a "p" orbital which is a dumbbell shape. The p-orbital has two lobes. Because the shape can be oriented in three different directions, there are three p orbitals pointing along the x, y and z axes and they are equal in energy (degenerate). After this we have the d-orbitals which have four lobes like a four leave clover. Finally the f-orbitals have 8 lobes that point to the corners of a cube.

The energy level of an orbital is determined by two factors - the size and shape. Usually size counts more, but there is one exception.

As each orbital becomes more complex, there are more ways to arrange their orientations, and therefore there are more orbitals available to hold electrons.

Finally, the more complex the orbital, the larger the shell must be for it to exist. p-orbitals start in the second shell, d-orbitals start in the third and so on.

The two big rules are "Pauli's exclusion Principle" and "Hund's rule of Maximum Multiplicity" - both can be explained very simply.

Pauli's exlcusion principle simply states there can only be two electrons per orbital and they must have opposite "spin" - one "up" and one "down" - we see this in the orbital diagrams, and it is the reason that there are only two electrons per orbital.

In each case the size of the orbital is related to the shell number.

Orbital	Typical Shape	Number of orbitals	Electrons per orbital	Max. electrons in sub-shell
s	Spherical	1	2	2
p	Dumbbell	3	2	6
d	4-leaf clover	5	2	10
f	Pointing to corners of a cube	7	2	14

Table 2.7: Orbital Properties

There are two factors which affect the energy level of an orbital; the size and shape. Usually bigger orbitals have a higher energy, but if an orbital has a complex shape, a smaller orbital can have a higher energy.

In the diagram, we can see the relative energies of the various orbitals.

As is evident by the periodic table, the 4s orbital is lower in energy than the 3d orbital, this means that electrons fill the 4s orbital first, then the 3d.

You can see this pattern in the periodic table. The periodic table is broken into sections or blocks - s, p, d and f. You can identify any element by its electron configuration by using the pattern below as a type of co-ordinate system. For example, bromine has the configuration ending in $4p^5$. 4 means the fourth shell/period/energy level; p, means the "p" block, and 5 means the fifth element in the block. Having identified the location of the element, you now just read off all the full sections that came before it.

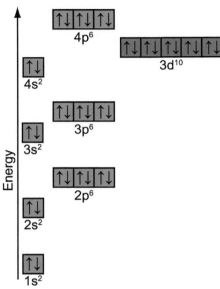

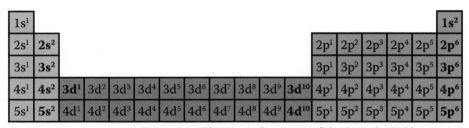

Figure 2.8: Electronic Structure of the Periodic Table

You can use the periodic table as a "map" of electron configurations, just read them off. For example, Bromine is $1s^22s^22p^63s^23p^64s^23d^{10}4p^5$. Just read across your table starting from hydrogen and mention each full orbital until you get to the last one which is the element you desire.

Electron configuration of ions. – just add or subtract electrons as appropriate.

Note

If the ion is a d-block element, it will lose the $4s^2$ electrons first, and then the 3d electrons after that. This is why all transition metals have a 2+ ion. See Chapter 3 - Transition Metals.

The electron configuration is determined by the Aufbau Principle - literally the building up of the electrons into the shells. In this way we move from the inner most orbitals to the outmost to build up the complete electron configuration of the atom. This idea is important, as in a few pages we will do the opposite process to strip the electons off of the atom in Successive Ionization energy.

It is possible to also have a shorter electron configuration. As the noble gases represent a completely full electron shell, we can start from there and continue building up until we arrive at the desired element.

P = $1s^2\,2s^2\,2p^6\,3s^2\,3p^3$

P = $[Ne]3s^23p^3$

In this case the first two full shells can be represented by the electron configuration of neon. It's important to note that this only applies to the noble gases. It would be incorrect to say $[Mg]3p^3$.

Atomic Structure

Here are the full ground state electron configurations for the following atoms & ions.

Cl	$1s^2 2s^2 2p^6 3s^2 3p^5$
Fe	$1s^2 2s^2 2p^6 3s^2 3p^6 4s^2 3d^6$
Se	$1s^2 2s^2 2p^6 3s^2 3p^6 4s^2 3d^{10} 4p^4$
O^{2-}	$1s^2 2s^2 2p^6$
Na^+	$1s^2 2s^2 2p^6$
Fe^{2+}	$1s^2 2s^2 2p^6 3s^2 3p^6 3d^6$ — note: loss of $4s^2$ electrons first
Fe^{3+}	$1s^2 2s^2 2p^6 3s^2 3p^6 3d^5$ — note: loss of $4s^2$ electrons + 1 "d" electron

Example

Write the full electron configuration for the following species.

N	
S	
Mg	
Al^{3+}	
Cl^-	
Cu^+	

2.4 Learning Check

AHL - Energy of Light

The energy of a light is related to its colour, where blue is more energetic than red. This principle covers the entire electromagnetic spectrum from radiowaves to gamma rays.

The energy is determined by the following formula (in your databook)

$$E = h\nu$$

where E is the energy in Joules, and ν is the frequency of the light in Hz (s^{-1}). The conversion factor, h, is known as Plank's constant has has a value of 6.63×10^{-34} J•s.

However, it is common in chemistry to speak of the wavelength of light instead of the frequency. In order to convert from frequency to wavelength we use the wave equation:

$$c = \nu \lambda$$

where c is the speed of light (3×10^8 ms^{-1}) and wavelength of visible light is typically between 400 and 700 nm (10^{-9} m) substituting in gives the resulting equation:

$$E = hc/\lambda$$

so, as the wavelength gets longer, the energy decreases.

The wavelength of the yellow line in the sodium emission spectrum is 589 nm. Determine the energy of this light.

Example

$E = hc/\lambda$
$E = 6.63 \times 10^{-34}$ J•s $\times 3 \times 10^8$ m•s^{-1} / 589×10^{-9} m
$E = 1.1256 \times 10^{-27}$ = 1.13×10^{-27}

1. Determine the energy of red (700 nm) and blue (400 nm) light.
2. What is the wavelength corresponding to 3.62×10^{-19} J?
3. Show that the ionization energy of hydrogen (1312 kJmol^{-1}) corresponds to 91 nm ultraviolet light.

2.5 Learning Check

Orbital Diagrams & Hund's Rule

Electron distribution can also be described by orbital diagrams, and these give another piece of information to the electron configuration.

Orbitals are represented by boxes, and electrons by up and down arrows so that each box has only one up and one down arrow. Notice that there are 3 orbitals in the p-sublevel.

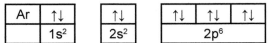

The "p" and "d" block orbitals are *degenerate* - that is they have equivalent energies for all three or five orbitals in the group. This is because they have the same size and shape, only different orientations.

Hund's Rule states that each orbital will fill singly first, then, when each orbital has one electron, they start doubling up. Because of this, the orbitals fill singly first of all (see carbon and nitrogen below), and then double up (see Oxygen)

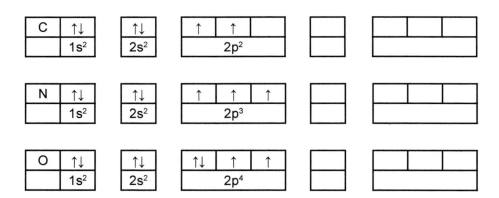

Exceptions to the filling pattern in the d-block.

The 4s and 3d subshells are very close in energy. Whenever a subshell is completely filled or exactly half filled, it is more stable (has a lower potential energy). This extra stability worth re-arranging some other electrons. Chromium and copper both have a d shell which is one electron short of being half or completely full. A 4s electron is moved to provide the extra electron to satisfy the half or full d-shell.

	Expected	Observed
Cr	$1s^22s^22p^63s^23p^64s^23d^4$	$1s^22s^22p^63s^23p^64s^13d^5$
Cu	$1s^22s^22p^63s^23p^64s^23d^9$	$1s^22s^22p^63s^23p^64s^13d^{10}$

Figure 2.9: Expected and Observed electron configurations

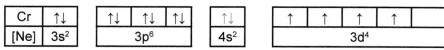

Figure 2.10: Expected orbital notation for Chromium

Figure 2.11: Observed orbital notation for Chromium

Successive Ionization Energies – evidence for shells

As successive electrons are removed from the atom, we see that there is a disproportionate jump as one of the electrons is removed. This jump in ionization energy represents the removal of an electron from a shell closer to the nucleus than the previous electrons removed.

Element	Ionization Eneries / kJ mol^{-1}							
	I_1	I_2	I_3	I_4	I_5	I_6	I_7	I_8
Na	496	**4562**	6910	9543	13354	16613	20117	25496
Mg	738	1451	**7733**	10543	13630	18020	21711	25661
Al	578	1817	2745	**11577**	14842	18379	23326	27465
Si	787	1577	3232	4356	**16091**	19805	23780	29287
P	1012	1907	2914	4964	6274	**21267**	25431	29872
S	999.6	2252	3357	4556	7004	8496	**27107**	31719
Cl	1255	2295	2850	5160	6560	9360	11000	**33604**
Ar	1527	2665	3945	5770	7230	8780	12000	13842

Table 2.12: Successive Ionization Energies

Look at silicon as an example; the first four ionization energies follow a trend, and the ionization energy takes a significant jump at the fifth one. This shows that the fifth electron is disproportionately harder to remove. We infer that the fifth electron is in a shell closer to the nucleus, therefore feels a stronger attraction and is harder to remove.

$$Si(g) \rightarrow Si^+(g) + e^- \qquad \Delta H_{first\ I.E} = 780 \text{ kJ} \cdot \text{mol}^{-1}$$
$$Si^+(g) \rightarrow Si^{2+}(g) + e^- \qquad \Delta H_{second\ I.E} = 1575 \text{ kJ} \cdot \text{mol}^{-1}$$
$$Si^{2+}(g) \rightarrow Si^{3+}(g) + e^- \qquad \Delta H_{third\ I.E} = 3220 \text{ kJ} \cdot \text{mol}^{-1}$$
$$Si^{3+}(g) \rightarrow Si^{4+}(g) + e^- \qquad \Delta H_{fourth\ I.E} = 4350 \text{ kJ} \cdot \text{mol}^{-1}$$
$$Si^{4+}(g) \rightarrow Si^{5+}(g) + e^- \qquad \Delta H_{fifth\ I.E} = 16100 \text{ kJ} \cdot \text{mol}^{-1}$$

We'll see more about ionization energy in Chapter 3.

Typically successive ionization energies are plotted using a logarithmic scale on the y-axis. This is due to the large scale involved in the graph. You should be able to sketch any graph or interpret a given graph. Remember that this graph is different from the trend in First Ionization Energy across Period 3 - many students get these two confused. In this case we "peel" the electrons off the atom from the outside in.

The scale is so large that it is more convenient to use the $\log_{10}$ of the ionization energy to compress the small differences and show only the larger jumps.

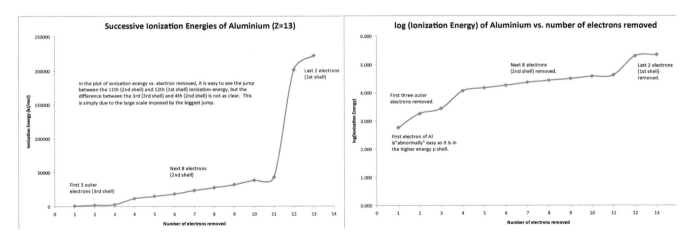

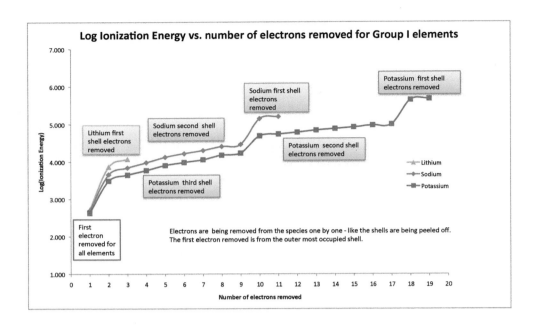

Exceptions to Ionization Energy Trend - Evidence for subshells

Before we spoke about the stability of half or full orbitals. Now we see this manifest in the ionization energy.

The **general** trend is for Ionization Energy to increase across the period.

Boron, and oxygen have ionization energies lower than beryllium and nitrogen respectively, breaking the trend. In the case of boron, you are starting a new p-obrital, and it's higher energy level means it is removed more easily.

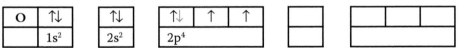

Oxygen has a 4th p-orbital electron, which is repelled by the electrons already in occupying each of the three p-orbitals.

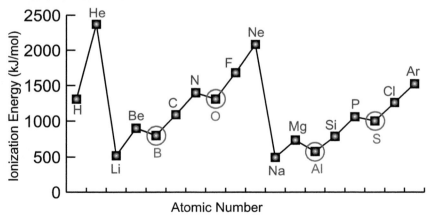

The same trend is seen in Period 3.

Figure 2.13: Exceptions to the general trend in Ionization Energy

Atomic Structure

Summary Questions

1. Determine the average atomic mass of element "Q" given the following data:
   ^{109}Q: 65.50%, ^{110}Q: 25.00%, ^{112}Q: 9.500%.

2. What is the natural abundance of each of ^{89}Lm and ^{90}Lm if the average relative mass is 89.65?

3. Write the electron configuration for the following species...

   a) P
   b) Se
   c) Br⁻
   d) Ca^{2+}
   e) Ni^{2+}

4. Draw orbital diagrams for the valence shell of...
   a) carbon
   b) sulphur

Chapter 3

Periodicity

In this chapter...

44	Structure of the Periodic Table
44	Effective Nuclear Charge
45	Trends in the Periodic Table - The Basic Idea
45	Atomic Radius
46	Ionic Radius
46	Atomic vs. Ionic Radii
48	Ionization Energy - General Trend
49	Electron Affinity
49	Electronegativity
50	Chemical Properties
50	Reactions of Alkali Metals and Halogens
50	Reactions of Alkali Metals with Water
50	Properties of the Halogens
51	Halogens & Halides
52	Chemical Properties across Period 3
52	Reactions of the oxides of Period 3 with water
53	First row d-block elements
53	Catalysts
54	Complex Ions
55	Colourful transition metal compounds
58	Summary Questions

Structure of the Periodic Table

The periodic table is simply a listing of the elements in order of atomic number (number of protons).

The vertical columns are called "groups" or "families", the horizontal rows are called periods. Groups have similar chemical properties. Groups are identified by their number 1-18. Older textbooks may still use a 1-8 system, but this is out-dated.

Periods represent the filling of a valence shell. Sodium, for example, has 3 shells, with one electron in the valence shell, whereas chlorine has 3 shells with 7 electrons in the valence shell.

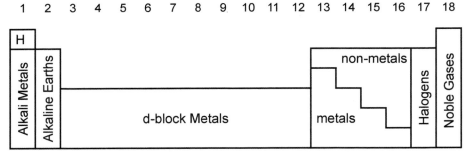

Figure 3.1: Structure of the Periodic Table

Physical Properties

In all the cases below, questions are designed to have **one variable change, and all others remain the same**. Your job in creating an answer is to correctly identify both of these. All of the physical properties depend ultimately on the balance of electrostatic attractions between oppositely charged particles (notice I didn't say protons & electrons) and the energy required to overcome these attractions. – Physics teachers will get on my case for comparing forces and energy.

In answering all these types of questions you must have a *logical sequence of thought*. The explanation must start with the physical reality, discuss the factors which are changing, and those which are staying the same. The changing factor must relate to a force of attraction between the relevant parts of the atom, and then finally how that force manifests itself with regards to the physical property under scrutiny.

Effective Nuclear Charge

Not examined by IBO, but it is valuable because it can help you understand the following properties. I'll explain it once here, and leave it to you to apply it later on.

The effective nuclear charge, Z_{eff}, arithmetically combines the positive charge of the nucleus and the negative charges of the core electrons (non-valence electrons).

Going down a group, the Z_{eff} is constant and going across a period the Z_{eff} increases.

eg. Magnesium has a Z_{eff} of 2+ due to 12 protons and 10 core electrons. In fact all of the group II metals have a 2+ Z_{eff}, so the changing radius is responsible for the trend it properties. Fluorine has a Z_{eff} of 7+ due to 9 protons, and only 2 inner electrons.

When working through the following trends, think about how the Z_{eff} is changing (or not).

Trends in the Periodic Table - The Basic Idea

All electrons, but most importantly the valence electrons, are held in orbit by the attractive force of the nucleus. In general, that force depends on two factors - the charge on the nucleus, and the square of the distance between the electrons and the nucleus.

$$\text{Force of attraction} = \frac{\text{nuclear charge}}{\text{distance}^2}$$

As you move across a period (in order of increasing atomic number), the number of protons in the nucleus increases, thus increasing the positive charge of the nucleus, which makes it more attractive to electrons. The increase of valence electrons does not contribute as much as the proton increase. Think about Z_{eff}.

As you move down a group, the number of electron shells increases, which increases the distance between the outer most (valence) electron and the nucleus, which decreases the force of attraction between the nucleus and the valence electrons. The addition of protons in the nucleus is cancelled out by the shielding of core electrons. (Again, you can think about Z_{eff} if you want.)

Atomic Radius

Atomic Radius is really quite straightforward – it's a measure of the distance from the nucleus to the outmost electrons. It may be measured in different ways (covalent radius, van der Waals' radius etc), but the idea is the same.

Once again, the forces within the atom are simple electrostatic – opposites attract and likes repel. So, the valence electrons are attracted by the nucleus and repelled by other electrons – both those in the same shell and those in the underlying shells.

As the atomic number increases across the periodic table (left to right), the only (significant) thing that is changing is the charge on the nucleus – there are more protons. The increased number of protons causes an increase in the attractive force between the outermost electron and the nucleus. With increased force comes decreased distance.

A common mistake is for students to think that the additional electrons make the atom get bigger, but they are filling the same shell, so are not contributing to the radius of the atom.

Below you can see the general trend of the radius decreasing along a period and increasing down a group.

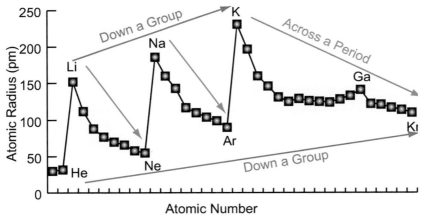

Figure 3.2: Atomic Radius vs. Atomic Number

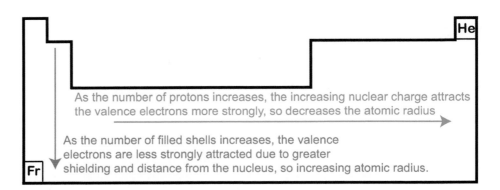

Ionic Radius

The definition is effectively the same as atomic radius, but now we are talking about ions.

Be careful to identify the electronic structure you are dealing with. Often you are asked to compare radii of positive and negative ions - sometimes with the same electron configuration, sometimes, with a whole shell difference!

Isoelectronic species

Isoelectronic just means particles with the same (exactly) electron number and therefore configuration. – This is actually quite easy, because 99% of the time, isoelectronic species are those ions that you probably learned before you started IB Chemistry. Isoelectronic species almost always have the noble gas electron configuration.

O^{2-}, F^-, Ne, Na^+, Mg^{2+} are isoelectronic – so what's the bit that's different? The protons of course – therefore Mg^{2+} with 12 protons is going have greater nuclear charge attracting the 10 electrons compared to O^{2-} with its 8 protons attracting 10 electrons.

	O	F	Ne	Na	Mg
protons	8	9	10	11	12
electrons	8	9	10	11	12
e- config.	$1s^22s^22p^4$	$1s^22s^22p^5$	$1s^22s^22p^6$	$1s^22s^22p^63s^1$	$1s^22s^22p^63s^2$
radius	66	58	54	186	160

Table 3.3: Atomic Radii of 5 elements

	O^{2-}	F^-	Ne	Na^+	Mg^{2+}
protons	8	9	10	11	12
electrons	10	10	10	10	10
e- config.	$1s^22s^22p^6$	$1s^22s^22p^6$	$1s^22s^22p^6$	$1s^22s^22p^6$	$1s^22s^22p^6$
radius	146	133	54	98	65

Table 3.4: Ionic Radii of 5 isoelectronic species

Atomic vs. Ionic Radii

So what happens when an atom becomes an ion?

Short answer – positive ions are smaller because the electrons occupying the outer shells are lost, negative ions are bigger because they gain repulsive forces.

3.1 Learning Check

Refer to your data book and see if you can justify the following ionic radii

Pair	Larger	Explanation
O / O^{2-}		

Periodicity

Pair	Larger	Explanation
Mg / Mg^{2+}		
Ne / F$^-$		
K / Ar		
K$^+$ / Ar		
Li / Na		
O / N		

Arrange the following groups in order of increasing size.
a) Be, Mg, Ca,
b) Te, I, Xe
c) Ga, Ge, In
d) As, N, F
e) S, Cl, F
f) Cs, Li, K

3.2 Learning Check

In each of the following sets, choose atom or ion that has the smallest radius? Justify your choices.
a) Li, Na, K
b) P, As
c) O$^+$, O, O$^-$
d) S, Cl, Kr
e) Pt, Pd, Ni
f) S^{2-}, Cl$^-$, Ar

3.3 Learning Check

Strange questions
Many times, you are asked questions which involve uncommon species as below. See if you can identify and explain the larger of the two species. This is best done by determining the relative number of protons and the electron configuration

Pair	Larger	Explanation
Mg$^+$ / Mg^{2+}		
O$^+$ / O$^-$		
Ar / Ar$^+$		
Na$^+$ / Mg$^+$		
Al^{2+} / Mg^{2+}		

3.4 Learning Check

Ionization Energy - General Trend

Definition

Minimum energy required to eject an electron out of a neutral atom or molecule in its ground state. (IUPAC)

$$M(g) \rightarrow M^+(g) + e^-$$

Usually the energy is measured for a mole of atoms, so that we can express the energy as kJ/mol.

Why neutral? – It means it's the first electron to come off from an **atom**, not an ion.

Note that the states are gaseous. That's because gaseous atoms are "infinitely" separated from their neighbours, and therefore do not have any confounding attractive forces – each atom is on its own.

The second ionization energy would be the energy required to remove the second electron from the +1 ion, and so on.

So what are the factors that affect ionization energy? – Simple – the force of attraction between the valence electron and the nucleus. The stronger the force, the more energy is required to pull it away.

Think about it as how much it costs to "buy" an electron. The more it's attracted to the nucleus, the more it costs.

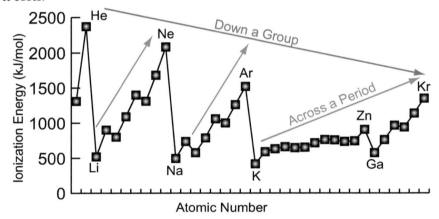

Figure 3.5: Trends in Ionization Energy

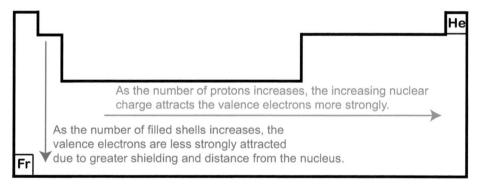

Figure 3.6: Trends in Ionization Energy

3.5 Learning Check

Arrange the atoms in Learning Check 3.2 in order of increasing first ionization energy.

A note about energy levels. – We speak of electrons in high energy levels.

Why is it that an electron in a higher energy level takes less energy to remove? Because it is closer to the "infinite" energy level (total removal). See Figure 2.

Electron Affinity

Electron Affinity: the energy released wehn an additional electron is attached to a neutral atom or molecule. (IUPAC)

$$X(g) + e^- \rightarrow X^-(g)$$

Or put more simply, this is the energy released when negative ions are formed.

The trend is the same – how much cash (energy) is an electron worth to an atom. Cheap electrons are found in France (Francium) – expensive electrons are found in the upper right corner of the periodic table – Fluorine.

Noble gases do not have an electron affinity because they already have a full outer shell and the nucleus is not strongly attractive enough to support attraction to the next shell.

Exception

Electronegativity

Electronegativity has several definitions, originally defined as the ability of an atom to attract electrons to itself, it now has different mathematical definitions. For our purposes we can say:

Electronegativity is the ability of an atom to attract a shared (bonding) pair of electrons.

Definition

The smaller Noble gases (He, Ne, Ar) are not assigned any electronegativity values because they do not form bonds.

This is really only useful later on for figuring out what type of bond you have. Large differences in electronegativity mean that the bond will be ionic; medium differences mean that it will be polar covalent, and small differences, covalent. See Table 4.1

The type of compound may also be measured by the average electronegativity. If the average electronegativity is low, then the substance is metallic (or an alloy). If the average is high, then the substance will be covalent - see your databook for the an Arkel-Ketelaar Triangle of Bonding.

Please note that electronegativity does not indicate bond strength - this is a common misunderstanding. Electronegativity helps to determine the type of bond, not how strong it is.

Note

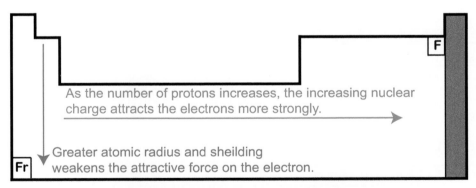

Figure 3.7: Trends in Electronegativity

Chemical Properties

Reactivity, is the ability of an *element* to form a compound or an ion.

Very reactive elements are found in nature as ions (eg. sodium), unreactive elements are found as the element (eg. gold).

Metals react by giving away electrons, forming positive ions called cations (*cat-eye-ons*). Non-metals react by taking electrons, and forming negative ions, anions.

Common Mistake

We don't generally speak of the reactivity of ions - they are the products of the reaction that has already occurred. Be sure to distinguish between the reaction that forms the ions, and the properties of the resulting ions.

The trends in metallic and non-metallic characteristics are based upon the physical properties

Reactions of Alkali Metals and Halogens

Alkali metals react with halogens to form their salts. Check you periodic table - the most vigorous reaction is between metals that lose electrons easily and halogens that are very good at taking electrons - eg. CsF.

$2Li(s) + Cl_2(g) \rightarrow 2LiCl(s)$	Increasing reactivity down the group, as metal atoms can lose their valence electron more easily.
$2Na(s) + Cl_2(g) \rightarrow 2NaCl(s)$	
$2K(s) + Cl_2(g) \rightarrow 2KCl(s)$	

Table 3.8: The formation of the chloride salts

As the metals become more reactive down the group (lower Ionization Energy) and the non-metals become more reactive going up a group (higher electron affinity), the most vigourous reaction occurs between cesium and fluorine.

As a side note, this is also the most ionic compound as the elements have the highest electronegativity difference.

Reactions of Alkali Metals with Water

Most students have seen or done these reactions. In all cases the product is a soluble base - called an alkali, and hydrogen gas.

$2Li(s) + 2H_2O(l) \rightarrow 2LiOH(aq) + H_2(g)$	slow reaction, bubbles produced	all produce gas, and move around on the water.
$2Na(s) + 2H_2O(l) \rightarrow 2NaOH(aq) + H_2(g)$	vigorous reaction, heat generated	
$2K(s) + 2H_2O(l) \rightarrow 2KOH(aq) + H_2(g)$	violent, hydrogen catches fire	

Table 3.9: The reactions of the alkali metals with water

Properties of the Halogens

Halogen	Colour in non-polar solvent	Colour in water
Cl_2	pale green/yellow	pale green/yellow
Br_2	reddish / orangy / brown	orangy/brown
I_2	purple / violet	dark straw (not very soluble)

Table 3.10: Identifying halogens by colour

Halogens & Halides

Remember that the **halogens** are F_2, Cl_2, Br_2, I_2 and the **halide** ions are F^-, Cl^-, Br^-, I^-. Many students mix them up. Be careful of the names!

Be careful!

Consider a halogen reacting with a halide ion. The ion has an extra electron, and the halogen has an affinity for electrons. Is the halogen strong enough to take the electron away from the halide ion? Maybe, it depends on their relative positions in the family.

Fluorine is the best electron taker, while iodine is the weakest. This is due to the fact that while all of the halogens have an effective nuclear charge of +7, fluorine has the smallest radius so the force of attraction is highest.

If you consider the reactions of bromine (Br_2), it is able to remove electrons from the "weaker" iodide ion (I^-), but not the chloride ion, Cl^-.

See Table 3.15 for the complete series including colour changes.

	Reaction	colour of non-polar layer
a)	$Cl_2(aq) + 2Br^-(aq) \rightarrow 2Cl^-(aq) + Br_2(aq)$	green to orange
b)	$Cl_2(aq) + 2I^-(aq) \rightarrow 2Cl^-(aq) + I_2(aq)$	green to purple
c)	$Br_2(aq) + 2Cl^-(aq) \rightarrow$ no reaction	stays orange
d)	$Br_2(aq) + 2I^-(aq) \rightarrow 2Br^-(aq) + I_2(aq)$	orange to purple
e)	$I_2(aq) + 2Cl^-(aq) \rightarrow$ no reaction	stays purple
f)	$I_2(aq) + 2Cl^-(aq) \rightarrow$ no reaction	stays purple

Table 3.11: Reactions demonstrating the relative reactivity of the halogens

Chemical Properties across Period 3

Exam Hint

There are a lot of reactions to memorize in this section. Many students attempt a Paper 2 question with good initial intentions of some easy periodicity, but then fade off quickly when they don't know the equations.

Reactions of the oxides of Period 3 with water

You are going to have to memorize the three * equations!

* $Na_2O + H_2O(l) \rightarrow 2NaOH(aq)$	product is very **basic**
* $MgO(s) + H_2O(l) \rightarrow Mg(OH)_2(s)$	product has low solubility & is **basic**
$Al_2O_3(s) + H_2O \rightarrow$ no reaction[1]	insoluble in water - see below
$SiO_2(s) + H_2O(l) \rightarrow$ no reaction[2]	insoluble in water - see below
$P_4O_6(s) + 6H_2O(l) \rightarrow 4H_3PO_3(aq)$	phosphorous **acid** formed
* $P_4O_{10}(s) + 6H_2O(l) \rightarrow 4H_3PO_4(aq)$	phosphoric **acid** formed
$SO_2(g) + H_2O(l) \rightarrow H_2SO_3(aq)$	sulphurous **acid** formed
$SO_3(g) + H_2O(l) \rightarrow H_2SO_4(aq)$	sulphuric **acid** formed
$Cl_2O(l) + H_2O(l) \rightarrow 2HClO(aq)$	hypochlorous **acid** formed
$Cl_2O_7(l) + H_2O(l) \rightarrow 2HClO_4(aq)$	perchloric **acid** formed

Table 3.12: Period Three oxide reactions
note the 3 reactions with * are required.

1) Aluminium oxide does not react easily with water, but does react with both acids and bases which makes it *amphoteric*.

$Al_2O_3(s) + 6H^+(aq) \rightarrow 2Al^{3+}(aq) + 3H_2O(l)$ dissolves in acid

$Al_2O_3(s) + +3H_2O + 2OH^-(aq) \rightarrow 2[Al(OH)_4]^-(aq)$ dissolves in base

Definition | A chemical species that behaves as both an acid and a base is called amphoteric (IUPAC)

2) It should be no surprise that silicon dioxide doesn't react with water - SiO_2 is the major component of glass. Silicon dioxide can react with a base to form sodium silicate, which is soluble. Therefore silicon dioxide is considered to be weakly acidic.

$SiO_2(s) + 2NaOH(aq) \rightarrow Na_2SiO_3(aq) + H_2O(g)$ dissolves in base

	Na_2O	MgO	Al_2O_3	SiO_2	P_2O_5	SO_3	Cl_2O_7
pH	13	11	7	6	1		
bond type	ionic		intermediate	giant covalent	molecular covalent		
solution	alkaline		amphoteric	insoluble	acidic		

Table 3.13: pH and explanations of period 3 oxides

Ionic oxides dissolve to produce alkaline solutions, whereas, molecular covalent oxides produce acid solutions.

As the ionic character decreases, and covalent character increases, the oxide of aluminium is intermediate between bonding types, and is amphoteric.

Periodicity

First row d-block elements

All transition elements are found in the d-block, but not all d-block elements are transition metals.

> A transition element is an element whose atom has an incomplete d sub-shell, or which can give rise to cations with an incomplete d sub-shell. (IUPAC)

Definition

Sc	Ti	V	Cr	Mn	Fe	Ni	Co	Cu	Zn
d-block elements									
transition elements									X

According to the definition, zinc is not a transition metal due to the fact that as a metal or an ion, it has a full d-shell; the Zn^{2+} ion is formed by loss of the $4s^2$ electrons.

When we talk of transition elements, we mean the elements that exhibit certain characteristic properties which are due to partially filled d-block orbitals.

Property	Reason
variable oxidation number (multiple ions)	loss of $4s^2$ electrons to give +2 ions and loss of other 3d electrons to give stable configurations - see your data book for all possible ions.
complex ion formation	partially filled d-orbitals (ability to act as a Lewis Acid)
colourful compounds	electron transitions between partially filled split d-orbitals.
catalytic properties	partially filled d-orbitals & variable oxidation states.
magnetic properties	unpaired electrons cause the material to be attracted to a magnetic field. (paramagnetism)

Table 3.14: Transition metal properties and reasons

Zinc does not have properties of transition metals. In compounds, zinc exists as a 2+ ion in its compounds (loss of the $4s^2$ electrons), and has a completely filled d-orbital.

Catalysts

> A catalyst is a species that increases the rate of reaction by providing an alternate pathway with lower activation energy.

Definition

Many transition metals are catalysts because of either their ability to form multiple oxidation states and/or the ability to provide partially filled d-orbitals to temporarily bond to a reactant species.

You should know the following examples of "Heterogeneous catalysts" - Catalysts in a different phase from the reacting species.

Get the point!

Process	Heterogeneous Catalyst
$2H_2O_2 \rightarrow H_2O(l) + O_2(g)$	MnO_2
$2SO_2(g) + O_2(g) \rightarrow 2SO_3(g)$ Contact Process	V_2O_5
$N_2(g) + 3H_2(g) \rightarrow 2NH_3(g)$ Haber Process	Fe
hydrogenation of alkenes	Ni
catalytic converters	Pd / Pt

Table 3.15: Catalysed reactions

Transition metals can also act as electron conveyors. Consider the Fenton reaction

Reaction 1: $H_2O_2 + Fe^{2+} \longrightarrow Fe^{3+} + HO\bullet + OH^-$

Reaction 2: $H_2O_2 + Fe^{3+} \longrightarrow Fe^{2+} + HOO\cdot + H^+$

The iron ions cycle back and forth from 2+ and 3+ while producing highly reactive HO• and HOO• radicals.

Complex Ions

Definition — A complex is a molecular entity formed by loose association involving two or more component molecular entities (ionic or uncharged)...The bonding is normally weaker than in a covalent bond. (IUPAC)
In this context IUPAC recommends the term "co-ordination entity"

A complex ion consists of a transition metal ion surrounded by ligands. In Figure 3.16, the silver ion is acting as a Lewis acid, and the two ammonia ligands are Lewis bases. The double headed arrows indicate that the co-ordinate bond is formed by donation of both electrons from the ammonia lone pair.

$$H_3N{:}\longrightarrow Ag^+ \longleftarrow {:}NH_3$$

$$[Ag(NH_3)_2]^+$$

Figure 3.16: Silver ammonia complex ion

A ligand is a ion or molecule containing lone pair of electrons that forms a **co-ordinate bond** to the central ion. A ligand is a Lewis base.

Remember that a co-ordinate bond occurs when one species (the ligand) donates both electrons to the bond. As it is a "loose association", one can change the ligands by increasing the concentration of the ligand in solution. This can be observed by adding concentrated ammonia to an aqueous solution of Cu^{2+} ions. The light blue caused by the H_2O ligands changes to dark purple when NH_3 ligands displace the water.

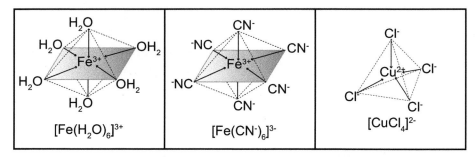

Figure 3.17: Some complex ions and their structures
(lone pairs not shown for clarity)

Colourful transition metal compounds

Compounds of transition metals are usually colourful. That means that there is an interaction with light. Some colours are absorbed, and some colours are not. So why are some wavelengths (or frequencies, or energies, or colours) absorbed?

In compounds, the presence of ligands causes some of the d-orbitals to be raised to a higher potential energy, and some are lowered. If we consider only the cases where there are 6 ligands (as in solutions where six water molecules hydrate the ion), the ligands will lie on the x, y and z axes. The electrons in the d-orbitals that are on the axes will be repelled more thatn the electrons that are in d-orbitals that lie between the axes. This causes the d-orbitals to have two different energy levels. They become "split".

Now remember that the transition metal behaviour rests upon the presence of at least one partially filled orbital. This means there is the possibility for a lower level electron absorb some energy and be promoted to a higher d-orbital energy level. The energy difference between split d-orbitals is the same amount of energy as visible light.

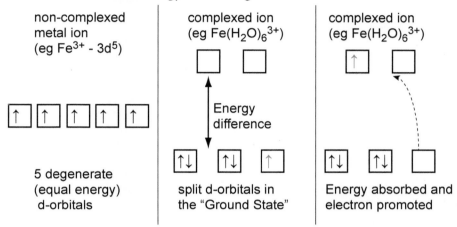

Figure 3.18: Splitting of d-orbitals into non-degenerate orbitals.

If the energy gap is large, then the compound will absorb a colour closer to the blue end of the spectrum. Small energy differences will cause red light to be absorbed.

Zinc compounds are not coloured because their d-block is totally full, and there is no place for the electron to be promoted into.

Be very careful to not confuse the absorption of light energy by promotion of electrons in transition metals with the emission spectrum of hydrogen. Both are due to electronic transitions, however transition metals do not emit light(i.e. glow in the dark), whereas the excited hydrogen electrons do.

The energy difference is a function of three factors. Greater nuclear charge will cause increased attraction to the nucleus and therefore lower the size of the gap. The number of electrons will affect the size of the gap based upon electron-electron interactions. And finally, the strength of the ligand will change the splitting. The strong-field ligands will cause a greater split than the weak ones. This is given in the data book and is referred to as the "Spectrochemical Series".

Summary Questions

1. In terms of electron configuration, outline the reasoning for the following observations...
    a) the first ionization energy of sulfur is less than phosphorus
    b) the first ionization energy of boron is greater than aluminium.
    c) the second ionization energy of sodium is greater than magnesium

2. State and explain the difference between...
    a) the atomic radius of nitrogen and oxygen.
    b) the atomic radius of nitrogen and phosphorus.
    c) the ionic radii of Si^{4+} and P^{3-}.

3. State and explain the trend in atomic radius and ionization energy of
    a) the alkali metals.
    b) the elements of period 3 Na to Ar.

4. Write balanced equations for the following reactions
    a) lithium and iodine
    b) potassium and water
    c) bromine and iodide

5. Speculate on the two likely oxidation states of titanium.
    a) Write their electron configurations.
    b) State with a reason, which one would you would expect to be colourful.

6. Explain why Cu^{2+} ions are colourful, but Zn^{2+} ions are not.

Chapter 4

Bonding

In this chapter...

Page	Topic
58	Bonding vs. Structure
58	The Octet Rule
58	Types of Bonding
60	Ionic Charges & Formulae
60	Simple ionic compounds – from the periodic table
60	Transition metal compounds
61	Is it +2 or 2+?
61	Compounds of polyatomic ions.
62	Ionic Compounds & Properties
62	Metallic Bonding & Alloys
62	Melting Point
62	Melting points of Group I – The alkali metals
63	Covalent Bonding
63	Covalent Formulae
64	Lewis Structure
65	VSEPR Theory and Molecular Shapes I
66	Exceptional Lewis Structures
67	Shapes of molecules II
68	Hybridization
69	Types of Covalent Bonds - Sigma(σ) & Pi(π)
70	Resonance & Delocalization
71	Resonance Stabilization Energy
71	Bond Angle and Double Bonds
72	Formal Charges
73	Bond Polarity and Molecular Polarity
73	Bond Length and Strength
74	Allotropes of Carbon
75	Intermolecular Forces
75	London Dispersion Forces
75	Dipole – Dipole Forces
76	Dipole - Induced Dipole
76	Hydrogen Bonding
77	The layering of intermolecular forces
77	How do you determine the type of IMF?
78	Types of Solids
78	Physical Properties of Solids
80	Summary Questions

Bonding vs. Structure

Many students have difficulty in distinguishing between these two separate, but related ideas.

The type of bonding is determined by the relative electronegativity of the two elements, whereas the structure is either "giant" or "molecular".

A molecule is simply a discrete unit of the compound. Molecules are only every found as a result of covalent bonding. One of the give aways to determine the type of structure is the melting point. Giant structures typically have high melting points because the lattice is held together with strong ionic, metallic or covalent bonds. Molecular structures have low melting points because the lattice is held together by the weak intermolecular forces between the molecules.

Bonding	Structure	examples	m.p.	lattice points
ionic	giant	$NaCl$, $MgBr_2$, $FeCl_3$	high	cations & anions
metallic	giant	Mg, Fe, K, Au	high	cations (with a sea of delocalized electrons
covalent	giant	diamond, graphite, Si, SiO_2*	high	atoms (localized bonds except graphite)
covalent	molecular	H_2O, CO_2, $C_6H_{12}O_6$	low	molecules with weak intermolecular forces maintaining lattice.

* There are very few examples of giant structures at the IB level, however ceramics would be a vast area of complex formulae that would also have giant structures.

The Octet Rule

You may have learnt the "Octet Rule", which may be stated something like: "atoms gain, lose or share electrons to end up with a complete octet in the outer shell". This works well for simple ion formation. For example oxygen gains two electrons to complete the outer shell, and aluminium will lose its three valence electrons. But for covalent compounds, sorry, but there are more exceptions to the rule than followers.

Firstly, hydrogen only requires two electrons in its outer shell because it only has access to the first shell.

Secondly, atoms in Group II or III typically form two or three bonds respectively, and are satisfied with fewer than eight electrons - these are often called electron deficient atoms.

Lastly, atoms with atomic numbers greater than silicon, can either follow the octet rule, or exceed it as they have access to more orbitals to create more bonds. These may be called "hypervalent" by some, but this assumes that the rule is to be followed.

So in summary, the octet rule is good for carbon, nitrogen, oxygen and fluorine - the Period 2 non-metals, and not much else.

Types of Bonding

In compounds there are really only two types of bonding – ionic and covalent. The problem is that it's not black and white – it's grey. The type of bonding depends on the difference in electronegativity. Large differences in electronegativity mean that the bond will be ionic; medium differences mean that it will be polar covalent, and small differences will be covalent. It is usual to speak of the "character" of the bond as being more ionic or more covalent.

The strength of the bond is determined by the degree of charge and the distance between charges, not electronegativity difference.

Bonding

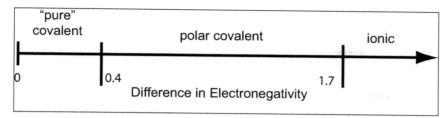

Figure 4.1: The Bonding Continuum

2.1																	He
1.0	1.5											2.0	2.5	3.0	3.5	4.0	Ne
0.9	1.2											1.5	1.8	2.1	2.5	3.0	Ar
0.8	1.0	1.3	1.5	1.6	1.6	1.5	1.8	1.8	1.8	1.9	1.6	1.6	1.8	2.0	2.4	2.8	Kr
0.8	0.9	1.2	1.4	1.6	1.8	1.9	2.2	2.2	2.2	1.9	1.7	1.7	1.8	1.9	2.1	2.5	Xe

Table 4.2: Values of Electronegativity in the Periodic Table

You don't always need to make the calculations, because you can generally make the following safe assumptions.

Metal − non-metal = ionic
Two different non metals = polar covalent
Two identical non-metals (Cl_2) = non-polar covalent bond (also C-H)

The greater the separation on the periodic table the greater the polarity and ionic character.

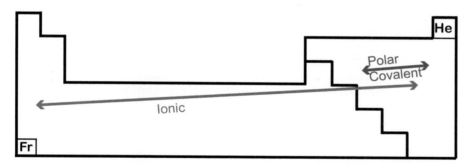

Figure 4.3: Relationship between separation and bond type

Ionic Bonding

An ionic bond is the electrostatic attraction between positive and negative ions. An ionic bond is often considered the result of the transfer of electrons, however this can cause confusion as students think about the reaction forming ions rather than the attraction between them after they are formed. A bond is ionic whene there is a large difference in electronegativity.

The positive ion is referred to as the "cation" and the negative ion is the "anion". This is a historical term derived from electrolysis and the ions that were found around the negative cathode were called cations, and anions were attracted to the positive anode.

From IUPAC:

> The bond between atoms with sharply different electronegativities. In strict terms, an ionic bond refers to the electrostatic attraction experienced between the electric charges of a cation and an anion, in contrast with a purely covalent bond. In practice, it is preferable to consider the amount of ionic character of a bond rather than referring to purely ionic or

purely covalent bonds.

The latter comment is quite important, because we no longer have purely ionic or purely covalent bonds (or compounds). We speak of how much ionic character a bond has. If it has more than 50% ionic character, we usually say it's ionic.

Ionic Charges & Formulae

Metals give away valence electrons to form positive ions, whereas non-metals gain electrons to complete the shell.

1+	2+
H	
Li	Be
Na	Mg
K	Ca

3+	4+/4-	3-	2-	1-	0
					He
B	C	N	O	F	Ne
Al	Si	P	S	Cl	Ar
Ga	Ge	As	Se	Br	Xe

Simple ionic compounds – from the periodic table

In order to write valid ionic formulae, there must be an overall charge of zero. All compounds are electrically neutral – the total of the positive charges must equal the total of the negative charges.

Consider the following compounds:

Compound	positive	negative	total +	total -	formula
lithium fluoride	Li^+	F^-	1+	1-	LiF
potassium oxide	K^+	O^{2-}	2(1+)	2-	K_2O
magnesium chloride	Mg^{2+}	Cl^-	2+	2(1-)	$MgCl_2$
calcium sulphide	Ca^{2+}	S^{2-}	2+	2-	CaS
aluminium chloride	Al^{3+}	Cl^-	3+	3(1-)	$AlCl_3$
aluminium oxide	Al^{3+}	O^{2-}	2(3+)	3(2-)	Al_2O_3

The formula of an ionic compound is simply the ratio of positive and negative the ions in the lattice - there is no such thing as "multiple bonds" in ionic bonding.

Each ion in the lattice is *co-ordinated* to several other ions depending on the lattice geometry.

4.1 Learning Check

Write the correct formulae for the following compounds.
a) sodium sulphide
b) beryllium fluoride
c) gallium iodide
d) potassium nitride
e) aluminium phosphide
f) magnesium nitride

Transition metal compounds

Transition metal compounds are even easier. Because there are multiple ions, you must be specific about which one you want. The ion (or really the oxidation state) is given in Roman numerals in brackets.

If you are asked to name a compound, you must use the negative charge to figure out the charge on the positive ion and consequently its name.

Iron(II) chloride Fe^{2+} Cl^- $FeCl_2$
Iron(III) chloride Fe^{3+} Cl^- $FeCl_3$

FeO O is 2- therefore iron must be 2+   iron(II) oxide
Fe_2O_3 negative charge is 6-, therefore EACH Fe is 3+;   iron(III) oxide.

Bonding

Element	Charge	Radius / pm	Melting point (K)
Li	+1	152	454
Na	+1	186	371
K	+1	231	337
Rb	+1	244	312
Cs	+1	262	302

Table 4.4: Melting Points of the Alkali Metals

You can get two points for one comment – "*charge density*". As the ion gets larger, the charge is diluted over the increased surface of the ionic sphere. As the charge density decreases, the force of attraction between the ion and the delocalized electrons is lowered, therefore the melting point lowers.

Covalent Bonding

Covalent bonding is the attractive force between two nuclei and shared pairs of electrons, due to overlap of their orbitals. It occurs when there is a high average electronegativity and a low difference in electronegativity between two atoms.

If one pair is shared, it is a single bond. Double bonds occur when two pairs of electrons are shared; triple is three pairs.

Like all other types of bonding, the size of the atom and charge of the nucleus and the number of electrons involved in the bond will be important factors.

Covalent bonding arises when there is a low difference of electronegativity, but a high average electronegativity - see the bonding triangle in your data book.

Covalent Formulae

The following molecules are common gases in chemistry questions.

Name	Formula	Lewis Structure	comment
oxygen	O_2	:Ö=Ö:	~20% of air
nitrogen	N_2	:N≡N:	~80% of air
carbon dioxide	CO_2	:Ö=C=Ö:	you breathe out, plants absorb
hydrogen cyanide	HCN	H—C≡N:	very poisonous
ethane	C_2H_6	(structure of ethane)	a flammable hydrocarbon gas
ethene	C_2H_4	(structure of ethene)	contains a double bond that is reactive
ethyne	C_2H_2	H—C≡C—H	contains a triple bond

Table 4.5: Some common covalent molecules

However, we need a way to draw structures for any molecule given, not just memorizing the common ones.

Lewis Structure

A Lewis structure shows all electrons either as a bonding pair or a lone pair (non-bonding pair). In order to draw a proper Lewis structure you must follow these steps...

- Count the total number of valence electrons.
- Adjust for charge (remember negative charges ADD electrons).
- Connect atoms to central atom with lines representing 2 electrons.
- Complete the octet on surrounding atoms (except hydrogen of course).
- Place extra electrons on the central atom.
- Ensure octet rule satisfaction by moving lone pairs to form double or triple bonds if necessary.
- Draw the molecule or ion as clearly as possible to show the shape.
- If you are drawing an ion, put the species in square brackets and indicate the charge outside the brackets.

Example — Draw the Lewis structure for CCl_4, NF_3, H_2O and CO_2.

	CCl_4	NF_3	H_2O	CO_2
# of electrons	4+4(7) = 32	5+3(7) = 26	3(1)+6 −1 = 8	4+2(6)=16
draw bond	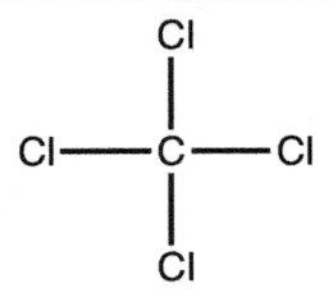 Cl—C(—Cl)(—Cl)—Cl	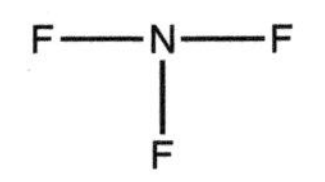 F—N(—F)—F	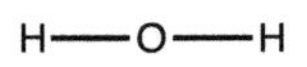 H—O—H	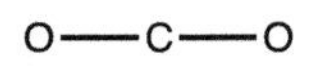 O—C—O
add lone pairs	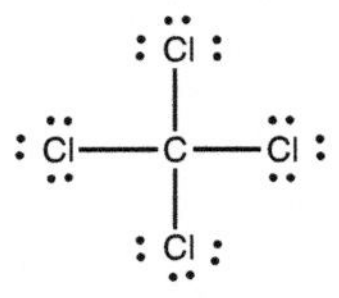 (with lone pairs on Cl)	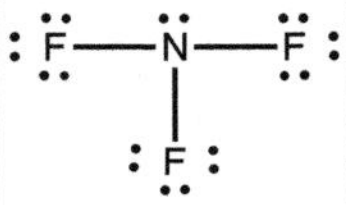 (with lone pairs on F)	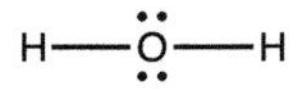 (lone pairs on O)	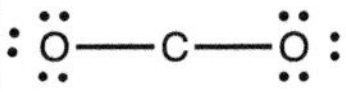 (lone pairs on O)
check octet	ok	ok	ok	C without 8e⁻
molecular drawing	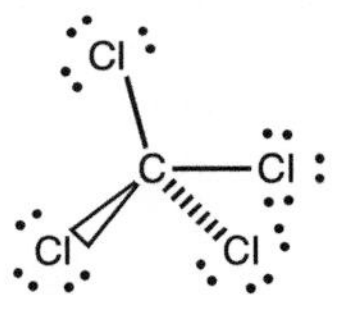 tetrahedral CCl_4	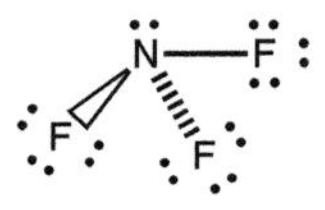 trigonal pyramidal NF_3	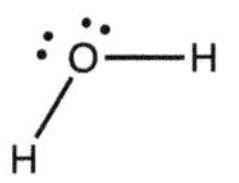 bent H_2O	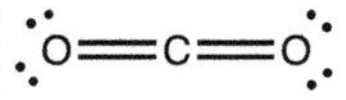 $\ddot{O}=C=\ddot{O}$

4.8 Learning Check

Draw Proper Lewis structures for each of the following species that obey the octet rule.
PH_3, $CHCl_3$, NH_4^+, H_2CO, SeF_2, PCl_4^+

VSEPR Theory and Molecular Shapes I

Valence Shell Electron Pair Repulsion (VSEPR) theory says that the pairs of electrons around the central atom will repel each other and therefore move to an orientation to minimize the repulsion and thus the lowest potential energy.

Lone pairs are attracted to only one nucleus, and therefore the orbital containing them will be short and fat, while orbitals containing bonding pairs are attracted to two nuclei will be drawn out. The net result is that lone pair lone pair repulsion is greatest, while bonding pair/bonding pair repulsion is least.

This means that the presence of lone pairs can disrupt the equal distribution of the electron domains around the central atom.

# of electron domains on central atom *	electron domain geometry	Bond Pairs	Lone Pairs	Molecular geometry (atoms only, not lone pairs)	Bond Angle	Example
2**	Linear	2	0	Linear	180°	:Cl—Be—Cl:
3**	Planar Triangle	3	0	Planar Triangle	120°	:Cl—B(Cl:)(Cl:)
4***	Tetrahedral	4	0	Tetrahedral	109.5°	CCl$_4$
4***	Tetrahedral	3	1	Trigonal Pyramid	107°	NCl$_3$
4***	Tetrahedral	2	2	Bent/ V-Shape	104.5°	OCl (with 2 lone pairs)

Table 4.6: Shapes predicted by VSEPR theory

* Electrons domains are either lone pairs, single bonds or multiple bonds - multiple bonds are still regarded as a single region of space where electrons are found.

** Beryllium and boron may form compounds with incomplete octets - just obey the formulae given to you. (i.e. "draw the structure of BCl_3"), and don't try to form double bonds. See Formal Charge.

*** This is the only group that actually follows the "octet rule". Not much of a rule is it?

You must distinguish between the electron domain geometry around the central atom and the molecular shape which only takes the atoms into account - not the lone pairs

The presence of a double bond in the Lewis structure does not affect the shape. The pi bond is not in the same plane as the other atoms, so no repulsion occurs within the plane of the sigma bonds. The double bond is still one electron domain

You need to only count the number of lone pairs and atoms surrounding the central atom.

Special Note

Name the shape in each case of Learning Check 4.8

4.9 Learning Check

Exceptional Lewis Structures

As previously mentioned there are several atoms that do not obey the octet rule.

Groups II & III atoms (Be, Mg, B, Ga etc) tend to be electron deficient - they are satisfied with less than 8 electrons.

Nonmetals-metals of Period 3 and higher (eg. sulphur) can exceed the octet rule because they have access to the 3d orbitals for types of hybridization that involve 5 or 6 orbitals. This type of hybridization is not a topic for IB Chemistry.

Sometimes you will come across compounds involving Noble gases. These compounds do exist, because they are the larger atoms combining with fluorine or oxygen, which have a very high electronegativity.

The trigonal bipyramidal shape has two different geometries. The "equatorial" and the "axial". There are three atoms or lone pairs on the equatorial plane, which are separated by 120°, and there are two atoms or lone pairs on the axis, which is at right angles to the equatorial plane.

Because lone pairs have greater repulsion than bonding pairs, the lone pair are placed in the equatorial positions where they are less crowded.

The octahedral shape is symmetric in all orientations, but after one lone pair is placed, the second lone pair must go in the position opposite.

The rules are the same otherwise.

4.10 Learning check

Draw Lewis structures for each of the following molecules that do not obey the octet rule for the central atom.

ClF_3, PF_5, SF_4, ClF_5, SbF_6^-, BF_3

Hint

These shapes are based on the "trigonal bipyramidal, octahedral and planar triangle" electron distributions.

4.11 Learning check

Use the table following to identify the shapes of the species in 4.10.

Shapes of molecules II

Total e- domains on central atom	Electron domain geometry	Bond Pairs	Lone Pairs	Molecular Geometry	Bond Angle	Molecular Shape Diagram
5	Trigonal Bipyramidal	5	0	Trigonal Bipyramidal	90° / 120°	
		4	1	See-Saw/ Distorted Tetrahedron	90° / <120°	
		3	2	T- Shape	90°	
		2	3	Linear	180°	
6	Octahedral	6	0	Octahedral	90° / 90°	
		5	1	Square Pyramid	90°/ 90°	
		4	2	Square Planar	90°	

Table 4.7: Shapes predicted by VSEPR that exceed the octet rule

Hybridization

Hybridization is the linear combination of atomic orbitals to create new orbitals with properties of those that were mixed.

Common Error

Don't mix up the definition of bonding, which is the **overlap** of orbitals. Hybridization is the **mixing**. You are a hybrid of your parents' genetic material.

There are 3 different ways to have hybridization occur in period 2 elements. There are others, but they are beyond the scope of the course.

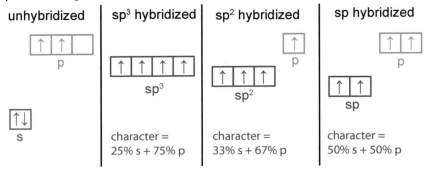

Figure 4.8: Hybrid Orbital energy levels

The first way, is the mixing of all 4 orbitals - one "s" and 3 "p" orbitals to create sp^3 hybrid orbitals.

The next way is to mix one s orbital and 2 p orbitals forming sp^2, leaving one unhybridized p orbital. The last way is to mix one s and one p, leaving two unhybridized p orbitals to have sp hybridization.

You need to remember the unhybridized orbitals are what lead to "pi" bonding.

The main thing that you have to do on the exams is to match or identify the hybridization with the type of bonding in the molecule.

hybridizaation	bonding	geometry	angle
sp^3	all single	tetrahedral	109.5°
sp^2	double bond	planar triangle	120°
sp	triple bond or 2 double bonds	linear	180°

Note that it doesn't really matter what type of atom you are talking about, it will probably be carbon, but it could be oxygen or nitrogen too!

4.12 Learning Check

Complete the following table for the atoms indicated in bold type. You should sketch the molecule in the margin.

molecule	hybridization	shape	angle
CH_3CH_3	sp^3	tetrahedral	109.5°
CH_2CH_2	sp^2	trigonal planar	120°
O_2	sp	linear	not relevant - 2 atoms
CH_3CHO	sp^2	trigonal planar	120°
CH_3COCH_3	sp^2	trigonal planar	120°
CH_3CN	sp	linear	not relevant - 2 atoms
N_2	sp	linear	not relevant - 2 atoms
CH_3NH_2	sp^3	tetrahedral	107° (one lone pair)

Identify the hybridization and bond angles in each of the indicated atoms.

4.13 Learning Check

Bonding

Molecular Structure	Hybridization	Bond Angle	Type of Bond	Shape
$_1$H$_3$C—C(H$_2$)$_2$—C(H$_2$)$_3$—C$_4$(H)=CH$_2$ $_5$	1 sp^3	109.5°	single	Tetrahedral
	2 sp^3	109.5°	single	Tetrahedral
	3 sp^3	109.5°	single	Tetrahedral
	4 sp^2	120°	double	Trigonal Planar
	5 sp^2	120°	double	Trigonal Planar
HC$_2$≡C$_3$=CH$_2$ $_4$ with H$_3$C $_1$	1			
	2			
	3			
	4			
$_1$H$_3$C—C$_2$≡C$_3$—CH$_3$ $_4$	1			
	2			
	3			
	4			
$_1$H$_3$C—C$_2$(=O $_4$)—CH$_3$ $_3$	1			
	2			
	3			
	4			
$_1$H$_3$C—C$_2$(=O $_4$)—OH $_3$	1			
	2			
	3			
	4			

Types of Covalent Bonds - Sigma(σ) & Pi(π)

A sigma(σ) bond is the end-on-end overlap of orbitals along the internuclear axis.

The electron density of the sigma bond exists along the internuclear axis (the imaginary line that connects the nuclei). The orbitals involved may be s, p, or hybrids sp, sp^2 or sp^3. Consider methane - the carbon's sp^3 hybrid orbitals overlap with the hydrogen's s orbital.

A pi(π) bond is the sideways overlap of unhybridized p-orbitals parallel to the internuclear axis

A single bond is only a sigma bond.
- A double bond is made of one sigma bond and one pi bond.
- A triple bond is made of one sigma bond and two pi bonds.

A sigma bond is not the overlap between "s" orbitals. - It may be, but is also the overlap between sp^3 hybrids, and s orbitals with sp^3 hybrids. In HCl, the hydrogen's s-orbital is sigma bonding with the sp^3 hybrid orbitals of the chlorine atom.   Common mistake

A pi bond is weaker than a sigma bond - check the data book!

Resonance & Delocalization

Resonance occurs when you can draw different valid Lewis structures for a single molecule. The molecule will have at least one double bond that may exist between different atoms. When the pi bond can "move around" the molecule or ion, it is said to "resonate" between the different structures. In fact the pi bond is distributed over all the atoms.

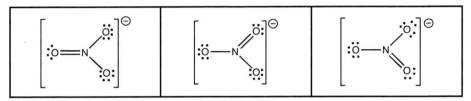

Figure 4.9: The three resonance structures of the nitrate ion.

In truth, the pi bond does not exist locally between any two atoms, but is delocalized.

Due to the presence of unhybridized p-orbitals on all of the atoms in the resonance structures, the pi bond is spread out over the whole molecule, or is delocalized.

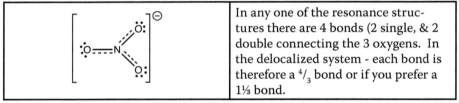

| | In any one of the resonance structures there are 4 bonds (2 single, & 2 double connecting the 3 oxygens. In the delocalized system - each bond is therefore a $^4/_3$ bond or if you prefer a 1⅓ bond. |

Figure 4.10: Delocalized bonding in the nitrate ion.

Common mistake Do not draw the fractional bonds as a valid Lewis structure as it is hard to account for the correct number of electrons

Figure 4.11: Benzene and nitrate's delocalized pi clouds

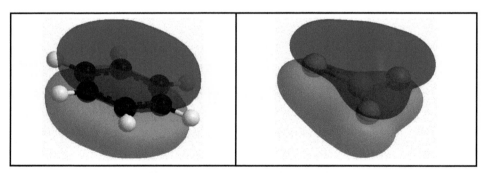

Resonance Structure 1	Resonance Structure 2	Delocalized Structure
Nitrite ion		
Ozone		

Bonding

[Benzene resonance structures]

Benzene

[Carboxylate ion resonance structures]

carboxylate ion

Note that the carboxylic acid group and it conjugate base have different bond lengths. In the acid, there exists a single and double bond, whereas in the carboxylate ion (above) resonance makes the two bonds equal length.

For each of the following, draw all possible resonance structures. SCN^-, SO_2, SO_3, CO_3^{2-}

4.14 Learning Check

Determine the bond orders for each species in Learning Check 4.14.

4.15 Learning Check

Resonance Stabilization Energy

Consider the theoretical compound cyclohexa-1,3,5-triene or 1,3,5-cyclohexatriene, where the electrons are localized, and we have three single and three double bonds vs. six intermediate bonds in benzene. We will ignore the C-H bonds, as they are the same in both compounds.

3 x 614 kJ/mol + 3x346kJ/mol = 2880 kJ/mol for the localized model

6 x 507 = 3042 kJ/mol for the delocalized model

This means that the delocalized model is 162 kJ/mol more stable than the localized model. This extra stability derived from the electrons ability to spread out around the ring is called resonance stabilization energy.

Bond Angle and Double Bonds

To be simple - double bonds don't affect the bond angle - that's why we speak of "electron domains" not pairs of electrons.

In order to make a double bond we have to hybridize $s+p_x+p_y$ orbitals to give 3 sp^2 orbitals who exist in the x-y plane. But remember the third p-orbital is responsible for the pi-bond and it lies in the z-axis. As the unhybridized p_z orbital is perpendicular to the hybrid orbitals it has no effect on the bond angles in the x-y plane.

Formal Charges

We can use an assignment of formal charges to determine the correct Lewis structure if we have two choices. The formal charge is a comparison of how many electrons an atom has access to compared to its normal valence. An atom "owns" its lone pairs, and shares its bonding pairs, so lone pairs count for two electrons, and bond count for only 1 electron.

The formal charge may be determined as follows:

Formal Charge (FC) = normal valence of atom - number of bonds to the atom - number of unshared electrons.

Formal Charge will help us identify the correct Lewis structures when we have electron deficient atoms, structures that exceed the octet rule and the correct resonance structure.

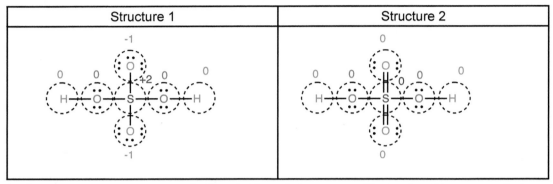

Looking at structure 1, we can calculate the formal charge on sulfur is +2 and is zero on all the oxygens and hydrogens. In structure 2 the formal charge on all atoms is now zero, making this the preferred Lewis structure. Recall that sulfur can exceed the octet rule as it has access to the d-orbitals.

This also works to explain the reason that group III atoms are electron deficient.

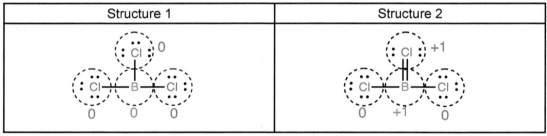

In structure 1, all atoms have a formal charge of zero. In structure 2, not only do the atoms have formal charges of +1 and -1, the more electronegative chlorine has the positive charge. Clearly we should use structure 1.

Consider the following possible structures that a student has drawn for nitric acid, HNO3:

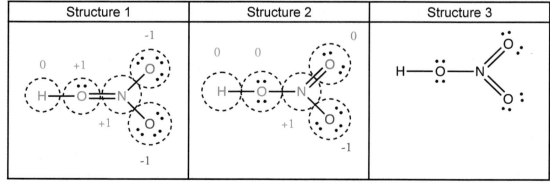

We can eliminate structure 3 straight away as nitrogen has 5 bonds, and this is not possible for period 2 elements. Structure 1 has a lot of formal charges, whereas structure 2 has only a +1 on nitrogen and a -1 on an oxygen. Structure 2 is favoured.

Bond Polarity and Molecular Polarity

A bond is polar when there is a difference in electronegativity between the two atoms bonded.

The more electronegative atom has a partial negative charge "∂^-", while the less electronegative atom has a partial positive charge, "∂^+".

A molecule will be polar if there is an overall polarity difference when you consider all the bonds, and their orientation in a molecule.

If the dipole moments (like vectors) cancel out, then there will be no net dipole and the molecule will be non-polar.

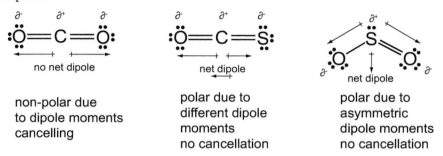

non-polar due to dipole moments cancelling

polar due to different dipole moments no cancellation

polar due to asymmetric dipole moments no cancellation

Figure 4.12: Bond Polarity and Dipole Moments

Property	Polar Molecule		Non-Polar Molecule
ΔE-neg	Significant difference in electronegativity		zero or very low difference in electronegativity
bond type	polar bonds	polar bonds	non-polar bonds
symmetry	non-symmetrical	symmetrical	doesn't matter
result	net dipole	dipoles cancel, no net dipole	no dipoles

Table 4.13: Properties of polar and non-polar molecules

Bond Length and Strength

As the number of shared electrons increase, the attractive force increases, so the bond strength (bond energy / bond enthalpy) increases. As the force of attraction increases, the bond length decreases.

	Number of bonds	Bond Length (pm)	Bond Strength (kJ/mol)
C–C (σ)	1	0.154	346
C=C ($\sigma + \pi$)	2	0.134	614
C≡C ($\sigma + \pi + \pi$)	3	0.120	839
C=C (benzene)	1.5	0.140	507

The strength of the sigma(σ) bond is 348 kJmol^{-1}.

The strength of a pi (π)bond is 612 − 348 = 264 kJmol^{-1} − weaker than a sigma bond, the double bond is not doubly strong, because the pi bond is weaker than the sigma bond. This is because the electron density of pi bond exists above and below the inter-nuclear axis.

Allotropes of Carbon

Definition: Allotropes are different molecular or crystalline structures of an element.

Carbon exists as three different allotropes described below.

	Diamond	**Graphite**	**Fullerene**
Hybridization	sp^3	sp^2	sp^2
Conductivity	None – all electrons are used in localized bonds	High – only 3 electrons used in bonding – the fourth electron is found in a delocalized π cloud	Intermediate – surface of the "ball" is conductive, but imagine an ant running across the surface of a lot of basketballs
Geometry	3-dimensional	2-dimensional	2 dimensional folded into a sphere
Hardness	Very hard – all atoms are held together in the lattice by strong covalent bonds	Very soft – 2 dimension sheets slide over each other because of weak van der Waals' forces between the sheets	Soft – imagine the molecular form of the "ball room at McDonald's"

Table 4.14: The properties of the allotropes of carbon

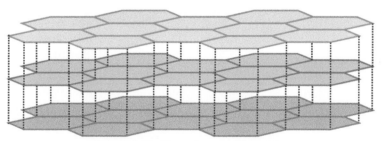

Figure 4.15: Three layers of Graphite

In Figure 4.14 each coloured sheet is attracted to the others by weak van der Waals' forces - shown by the dotted lines. The electrical conductivity occurs along a single sheet due to the delocalized, unhybridized p-orbital cloud above and below each plane of hexagons.

Many students get confused between the two properties...

Property	Explanation
Conductivity	Each carbon in graphite is bonded to 3 others, therefore there is an extra unhybridized p electron which is able to form a delocalized pi bond across the entire sheet.
soft powder	Because each sheet is non-polar, the sheets have only van der Waals' forces acting between them, therefore they slide over each other very easily

Intermolecular Forces

Intermolecular forces are the forces that attract one molecule to another molecule. These are the forces that govern physical properties – melting & boiling points, density, volatility etc. There are three inter**molecular** forces, collectively called "van der Waals' forces".

The boiling point of a compound is a good measure of the strength of the van der Waals' forces because boiling is the process of separating the molecules from each other.

London Dispersion Forces

London Dispersion forces are the result of a temporary, instantaneous or momentary dipole inducing a dipole in a nearby particle and the resulting attraction between the two temporary dipoles. Because of the temporary nature of the attraction, Van der Waals' forces are relatively weak, and boiling points are relatively low.

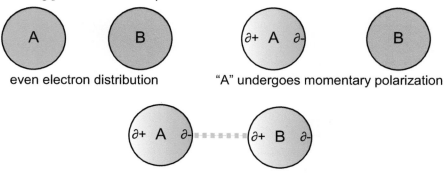

Figure 4.16: Van der Waals' forces

Dipole – Dipole Forces

This is perhaps the most common intermolecular force. It occurs between all polar molecules, which have permanent dipoles.

Because the dipole is now permanent, these forces are stronger than van der Waals' forces. Greater bond polarity leads to higher boiling points.

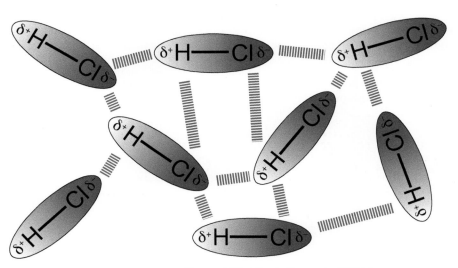

Figure 4.17: Dipole Forces in HCl(g)

Dipole - Induced Dipole

Dipole forces can also induce a dipole in a non-polar molecule.

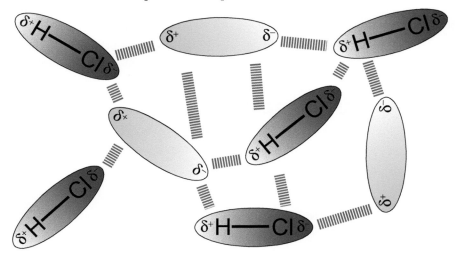

Figure 4.18: Interaction between dipoles and non-polar molecules with induced dipoles.

Hydrogen Bonding

There are only three structures that lead to hydrogen bonding. A hydrogen atom must be attached **directly** to a fluorine, oxygen or nitrogen (Hydrogen bonding is "FON" – like fun)

Figure 4.19: Hydrogen bonding between water and ethanol molecules

Exam Trap

Ethers (R-CH$_2$-O-CH$_2$-R), aldehydes(R-CHO), ketones(RCOR'), and esters(R-COO-R') do not have hydrogen bonding, because the H atom is not attached directly to an O atom. See Organic Chemistry for structures.

In some cases, hydrogen bonding gives rise to molecules "pairing up" to become dimers. The reciprocal hydrogen bonding in ethanoic acid is an example. As a consequence, it can appear to have double the formula weight.

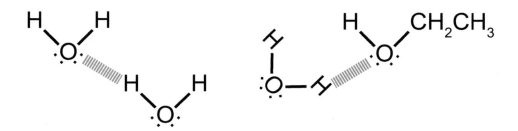

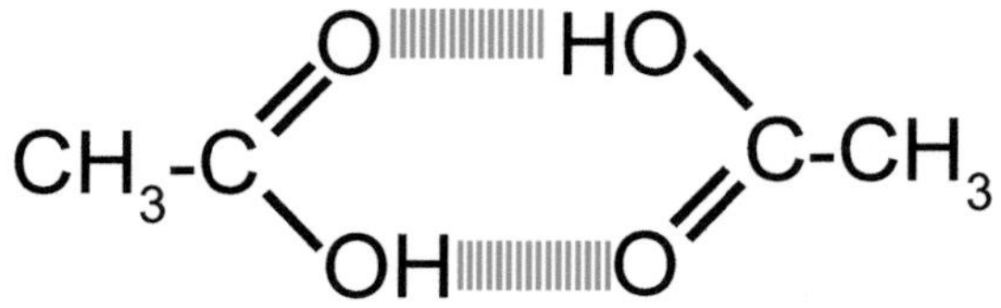

Figure 4.20: Hydrogen bonded dimer of ethanoic acid

Often we are concerned with the intermolecular forces that exists within a pure substance. We should also consider how a substance interacts in a mixture.

Propanone, without a O-H group, cannot hydrogen bond with itself, it can only use it's dipole-

dipole forces. So the boiling point of propanone is rather low at 56.5°C.

However, water can hydrogen bond to propanone's lone pairs on the oxygen. So propanone is soluble in water.

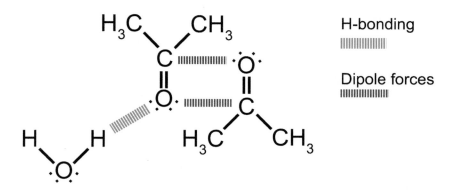

Figure 4.21: Propanone hydrogen bonding with water and the weak dipole-dipole forces between propanone molecules

The layering of intermolecular forces

It is common to speak of the strongest type of intermolecular force present. You must remember, however, that all of the weaker forces exist as well.

	H_2O	CH_3Cl	CH_4
strong	Hydrogen bond		
medium	dipole/dipole	dipole/dipole	
weak	London Forces	London Forces	London Forces

Table 4.22: The layering of Intermolecular forces

The comparisons assume that all other factors are reasonably equal - ie molecular mass. High molecular mass wax (a hydrocarbon with London forces) has a higher melting point than water.

How do you determine the type of IMF?

Is there a H atom attached directly to F, O or N? If "Yes" = Hydrogen bonding

Is the molecule non-polar (low Δ E-neg or symmetrical)? = Yes = London Forces

All others are dipole – dipole.

London Forces	Hydrogen bonding	Dipole – Dipole
Non-polar molecules e.g.... Any molecular element like Cl_2, Br_2, O_2, P_4, S_8 etc. and hydrocarbons CH_4 etc	H–F (only one molecule!) H–O– (alcohols, carboxylic acids, water) H–N< (amines, amides, ammonia)	Any other molecule that is not covered by the previous two. A covalent molecule that has different non-metals, not symmetrically oriented.

Table 4.23: Types of Intermolecular Forces

Types of Solids

In all types of bonding, we can consider a solid as a lattice of particles. The type of bonding depends on the type of particle and the forces that exist between them. It is important to distinguish between the structure (Giant vs. Molecular) and the type of bonding (Ionic, Covalent or Metallic)

Ionic - Giant Structure
The attractive force between oppositely charged ions. The lattice is made of oppositely charged ions. Salts are ionic. SALTS ARE NOT MOLECULES.

Covalent – Giant Structure
Regular covalent bonds, but they exist between ALL atoms – that is every atom is covalently bonded to all of its nearest neighbours. The lattice is made of atoms covalently bonded to each other.

Covalent – Molecular
Regular covalent bonds, but the atoms form little packages called molecules. Each package is held together by strong bonds, but they are attracted to each other by weak forces – see IMF below.

Metallic - Giant Structure
Metallic bonding is the electrostatic attraction between a lattice of positive ions and delocalized electrons.

Physical Properties of Solids

The strength of the forces between particles in the solid and liquid phases will determine the physical properties. In all cases there - especially molecular structures, we are not decomposing the compound, but only separating particles from each other.

See the short discussion on Melting Points in the Periodicity chapter.

Bonding	Lattice Forces	boiling point	volatility	ΔH_{vap}
(giant) ionic	Ion / ion attraction	high	low	high
(giant) metallic	ion / electron attraction	high	low	high
giant covalent	covalent bonds	high	low	high
molecular covalent	Hydrogen bonding Dipole / Dipole van der Waals'	low	high	low

Definitions

Volatility is a measure of the tendency for a substance to evaporate.

ΔH_{vap} is the potential energy change required to vaporize a liquid.

Summary Questions

1. Write the names of the following compounds...
   a) $Ca(NO_3)_2$
   b) $Cu(NO_3)_2$
   c) $Al_2(SO_4)_3$
   d) $Fe_3(PO_4)_2$
   e) NH_4Cl
   f) $NaHCO_3$

2. Determine the correct formula of the following compounds.
   a) potassium oxide
   b) ammonium nitrate
   c) magnesium carbonate
   d) nickel(III) oxide
   e) lead(IV) oxide
   f) aluminium phosphate

3. Draw a correct Lewis structure for the following covalent compounds...
   a) CO
   b) NI_3
   c) CH_2Cl_2
   d) NO_2^-
   e) NO_2^+
   f) KrF_2
   g) IF_3
   h) SeI_4
   i) AsF_5
   j) SF_6

4. Determine the dominant intermolecular force in the following molecules...
   a) CH_3OH
   b) CH_2Cl_2
   c) CH_3OCH_3
   d) CH_3CHO
   e) CH_3COOH
   f) CH_3COCH_3

5. Draw the three resonance structures of OCN^-.

6. State and explain which of the BCl_3 and NCl_3 has a higher melting point.

7. Discuss the number and types of bonding in propene. (points: σ / π, geometry and hybridization)

Chapter 5

Energetics

In this chapter...

- 82　Endothermic & Exothermic
- 83　Potential Energy Diagrams (Enthalpy Diagrams)
- 83　Standard Conditions
- 84　Calorimetry - Enthalpy Change Calculations
- 85　Assumptions in Calorimetry
- 86　The Calorimeter Constant
- 86　Hess' Law
- 88　Heat of Hydration - an application of Hess' Law
- 89　Bond Enthalpy
- 90　Standard Enthalpy Changes
- 91　Standard Enthalpy of Formation
- 92　Standard Enthalpy of Combustion
- 92　Which formula do I use?
- 93　Born – Haber Cycles & Ionic Lattice Enthalpy
- 94　Entropy
- 95　Hess' Law and Entropy
- 95　Spontaneity & Gibb's Free Energy
- 96　Free Energy & Equilibrium
- 97　Summary Questions

Endothermic & Exothermic

The first law of thermodynamics (Energetics) says that energy cannot be created nor destroyed; it can only change forms.

We divide the universe into two parts – the system and the surroundings and determine the energy flow between them. Most people have a basic understanding of what these two are, but there are some important details.

Heat is the energy transferred from a hotter body to a colder body due to a temperature gradient.
The temperature is a measure of the average kinetic energy of the particles. If the particles are moving faster, a higher temperature is observed.
The temperature at which all particles are stationary is called absolute zero. Zero Kelvin

As far as Chemists are concerned, the system is represented by the potential energy of the bonds, and the surroundings are represented by the kinetic energy of particles. We cannot measure the potential energy directly in the lab, so we must react chemicals together and measure the change of the kinetic energy of the surroundings – and assume that they are equal (i.e. no energy escaped our measurement)

The chemical potential energy stored in the system is called "enthalpy" and is measured in kJ mol^{-1}/

So if we consider two chemicals reacting in aqueous solution – say an acid and a base reacting exothermically. The system made up of the acid (H^+ ions), the base (OH^- ions) and the water product (H_2O). The separate ions have a higher potential energy than the product, so the energy flows into the surroundings as heat, and we see the temperature of the surroundings increase.

If thermal energy is flowing into the system, then the surroundings are losing energy, and the system is gaining energy

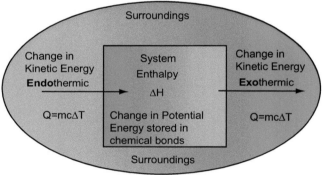

Figure 5.1: Energy flow in "The Universe"

Many students make the mistake of thinking that the solvent in an aqueous reaction is part of the system - but is it? - NO - the water that ions or molecules are dissolved in is not involved in the chemical changes. The solvent is part of the surroundings.

You need to think about the flow of energy (heat) between the system and the surroundings. If one gains, then the other loses. What form does the energy take in the different parts of "the universe"?

Property	Endothermic	Exothermic
Surroundings	Loses kinetic energy	Gains kinetic energy
System	Gains potential energy	Loses potential energy
Enthalpy change (ΔH)	Positive ($\Delta H > 0$)	Negative ($\Delta H < 0$)
Temperature change (ΔT)	Negative ($\Delta T < 0$)	Positive ($\Delta T > 0$)
Bonding	Bond Breaking	Bond Making

Table 5.2: Endothermic vs. Exothermic

As chemists, we are interested in the system - the chemical bonds. However we are forced to use changes in the surroundings to infer what happens in the system.

Potential Energy Diagrams (Enthalpy Diagrams)

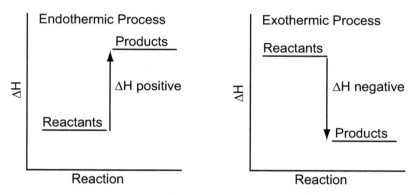

Figure 5.3: Potential Energy Diagrams / Enthalpy Diagrams

In the endothermic graph on the left, the system gains potential energy as the reactants become products. The value for ΔH is greater than zero.

In the exothermic graph on the right, the system loses potential energy to the surroundings as the reaction proceeds. The enthalpy change is negative.

In terms of stability, do not say that something is "unstable". Stability is relative. In Figure 5.3, the endothermic process has reactants which are more stable than products. In the exothermic process, the products are more stable.

Standard Conditions

Standard conditions are simply an agreed upon set of reference conditions under which to make the measurements.

There is Standard Temperature and Pressure (STP), which is 0°C / 273K and 100000 Pa (100 kPa), and there is Standard Ambient Temperature and Pressure (SATP) which is 25°C/298 K and 100000 Pa (100 kPa).

Some older authorities called SATP "room temperature and pressure (RTP), but this is not official.

The Standard values in this unit are noted by a ° symbol, so that ΔH° means the Standard Enthaply Change under SATP conditions (298K & 100 kPa).

The standard state of an element or compound is simply the physical state (solid, liquid, gas) of the element or compound under SATP conditions.

Standard Temperature and Pressure (STP) is mostly only used with gases and is not so common any more.

Also note that some older texts (and teachers) use 101.325 as standard temperature, but this is no longer the case (IUPAC changed it in 1982).

Calorimetry - Enthalpy Change Calculations

We are interested in how much energy change is related to 1.0 mol of products or reactants. This means that we want the energy per mole - typically it will be measured in kJ mol^{-1}. However, we only can measure the temperature changes.

In order to calculate a change in the potential energy of the system, we must measure the equal but opposite energy change in the surroundings. The energy change of the surroundings is given by

$$Q = mc\Delta T,$$

Where Q is the Quantity of energy in Joules

m is the mass of the material changing temperature

c is the specific heat capacity of the substance changing temperature

ΔT is the change in temperature ($T_2 - T_1$)

Physics students will be familiar with the specific heat capacity of water being 4200 J•kg^{-1}•C^{-1} – because physicists like kilograms. Chemists tend to prefer grams, but you shouldn't be thrown here.

Usually you have two masses – the mass of the material changing temperature, and the mass of the reactants (fuel) causing the change. – Don't get them mixed up.

Be Careful! Exam Trap

Because energy cannot be created or destroyed,

$$E_{lost} = -E_{gained}$$

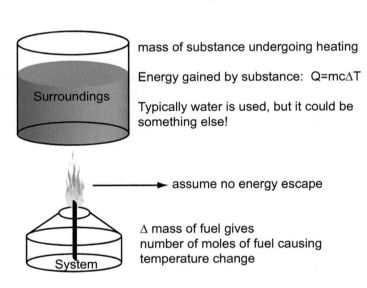

Figure 5.4: Determining the Enthalpy Change of a Fuel

As chemists we are interested in the energy change per amount of substance (moles) causing the change.

$$\Delta H = \frac{-Q}{1000 \cdot n}$$

Expression

Where ΔH is the enthalpy change of the system in kJ•mol^{-1},

Q is the energy change of the surroundings in J

n is the number of moles of **limiting** reactant.

we need to divide by 1000 to convert Joules to kilojoules.

Energetics

However, there are many cases where you aren't burning a fuel and capturing the heat. Often times you are mixing reactants. How will the change of volume or concentration affect the temperature change?

Mixing 50 cm³ of 0.1 M HCl and 50 cm³ of 0.1 M NaOH causes a temperature change of ΔT.

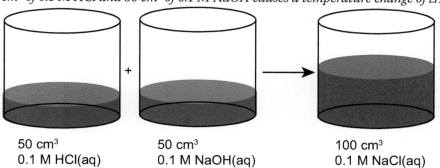

50 cm³
0.1 M HCl(aq)

50 cm³
0.1 M NaOH(aq)

100 cm³
0.1 M NaCl(aq)

Classic Question

a) How does mixing 100 cm³ of each solution affect ΔT?
b) How does mixing 100 cm³ of HCl(aq) with 50 cm³ of NaOH affect ΔT?
c) How does mixing 50 cm³ of 0.2 M solutions affect ΔT?

Answers
a) No Change! - yes you have doubled the heat, but it has to affect double the mass, so the two doublings cancel out.
b) Don't forget stoichiometry! - NaOH is still the Limiting reactant - therefore no change in the amount of heat produced, but it has to cause a temperature change for a larger mass, so the change in temperature is actually less!
c) Temperature change doubles. Doubling the number of moles without changing the mass is the only way to double the ΔT!

Assumptions in Calorimetry

A classic experiment is to react a known volume of exact concentration of copper(II) ions with an excess of zinc powder. This is an exothermic reaction and you can calculate the energy of reaction.

If 25.00 cm³ of CuSO4(aq) is reacted with 6.00 g of Zn(s), which represents an excess, in an insulated container (calorimeter), the temperature changes from 22 °C to 67 °C. Determine the ΔH of the reaction.

Classic Question

$Q = mc\Delta T$
$Q = 25.00 \text{ g} \times 4.18 \text{ J g}^{-1} \text{ °C}^{-1} \times (67.0 \text{ °C} - 22.0 \text{ °C})$
$Q = 4702.5 \text{ J}$
$Q = 4.70 \text{ kJ}$ - this is the energy gained by the water in the solution

$\Delta H = -4.70 \text{ kJ} \div 0.025 \text{ mol}$
$\Delta H = -188 \text{ kJmol}^{-1}$

The actual value is -217 kJ mol⁻¹. However we made some assumptions:

1. The specific heat capacity of the solution is the same as pure water;
2. The zinc / copper metal in the reaction mixture did not absorb some of the energy.
3. The density of solution was 1.0 g/cm³ (if we measured the volume of solution instead of the mass).
4. The calorimeter itself did not absorb any heat
5. Of course the biggest error factor in any energetics experiment will be heat loss.

The Calorimeter Constant

In many calorimeters, there are many components which may absorb heat - glass, metal etc. This introduces a problem in calculating the energy absorbed by each component. The solution to this problem is to determine how all of these absorb as a system, and then generate a constant.

For example a food calorimeter may have be comprised of a glass container with a copper chimney and hold 600 cm³ of water. We would have to do Q=mcΔT for each component, but if we combust some known compound, say 1. 5 g of ethanol, we can determine the ratio of the energy of combustion of the ethanol to the temperature change of the calorimeter, and call this ratio the calorimeter constant - it only works for that particular calorimeter, if it is filled with 600 cm³ of water, but it's easier.

Hess' Law

Enthalpy is what is known as a "state function", which means that it depends only upon the initial and final states, not the pathway between the states – the pathway is what is of concern in kinetics.

There are many ways to solve these problems. The simplest and most familiar way to solve these problems is to treat them as a system of equations.

You are always given equations to manipulate and a equation which is the goal.

Consider the following example.

Classic Question

Determine the enthalpy of reaction for

$$NO(g) + \tfrac{1}{2}O_2(g) \rightarrow NO_2(g)$$

given

$$\tfrac{1}{2}N_2(g) + O_2(g) \rightarrow NO_2(g) \qquad \Delta H = +33.8 \text{ kJ}$$
$$\tfrac{1}{2}N_2(g) + \tfrac{1}{2}O_2(g) \rightarrow NO(g) \qquad \Delta H = -113.14 \text{ kJ}$$

Looking at the first equation, we see our desired product, $NO_2(g)$ in the correct position, however in the second equation, the reactant NO is on the wrong side, so we will reverse the entire equation and change the sign of the enthalpy change to positive.

$$\tfrac{1}{2}N_2(g) + O_2(g) \rightarrow NO_2(g) \qquad \Delta H = +33.8 \text{ kJ}$$
$$NO(g) \rightarrow \tfrac{1}{2}N_2(g) + \tfrac{1}{2}O_2(g) \qquad \Delta H = +113.14 \text{ kJ}$$

This also allows us to cancel out the $N_2(g)$, which is not desired in our final equation. The oxygen cancels out to leave a half a mole on the left side.
Adding up the two equations, and the values for ΔH we get...

$$NO(g) + \tfrac{1}{2}O_2(g) \rightarrow NO_2(g) \qquad \Delta H = +146.94 \text{ kJ}$$

Remember, what ever you do to the equations, you do to the value of the enthalpy. It will and must work out to the final answer.

You will usually either flip the equation or double it or both.

Energetics

Now try the following...

5.1 Learning Check

1. Determine the heat of reaction for ...

$$BaO(s) + H_2SO_4(l) \rightarrow BaSO_4(s) + H_2O(l)$$

given

$BaO(s) + SO_3(g) \rightarrow BaSO_4$	$\Delta H = -213.0$ kJ
$SO_3(g) + H_2O \rightarrow H_2SO_4(l)$	$\Delta H = -78.2$ kJ

2. Determine the enthalpy for the following reaction

$$2NO(g) + O_2(g) \rightarrow N_2O_4(g)$$

given that

$N_2O_4(g) \rightarrow 2NO_2(g)$	$\Delta H = +57.93$ kJ
$NO(g) + \tfrac{1}{2}O_2(g) \rightarrow NO_2(g)$	$\Delta H = -56.57$ kJ

3. Hydrogen peroxide decomposes by the reaction...

$$H_2O_2(l) \rightarrow H_2O(l) + \tfrac{1}{2}O_2(g)$$

Determine the enthalpy of decomposition by manipulating the following equations.

$H_2(g) + O_2(g) \rightarrow H_2O_2(l)$	$\Delta H = -188$ kJ
$H_2(g) + \tfrac{1}{2}O_2 \rightarrow H_2O(l)$	$\Delta H = -286$ kJ

4. Given the following data:

$S(s) + \tfrac{3}{2}O_2(g) \rightarrow SO_3(g)$	$\Delta H = -395.2$ kJ
$2SO_2(g) + O_2(g) \rightarrow 2SO_3(g)$	$\Delta H = -198.2$ kJ

Calculate the ΔH of the reaction...

$$S(s) + O_2(g) \rightarrow SO_2(g)$$

5. Given the following data:

$C_2H_2(g) + \tfrac{5}{2}O_2(g) \rightarrow 2CO_2(g) + H_2O(l)$	$\Delta H = -1300$ kJ
$C(s) + O_2(g) \rightarrow CO_2(g)$	$\Delta H = -394$ kJ
$H_2(g) + \tfrac{1}{2}O_2(g) \rightarrow H_2O(l)$	$\Delta H = -286$ kJ

Calculate ΔH for the reaction...

$$2C(s) + H_2(g) \rightarrow C_2H_2(g)$$

C.Lumsden   IB HL Chemistry

Heat of Hydration - an application of Hess' Law

As stated above, one of the advantages of Hess' Law is to be able to determine the energy change for inconvenient reactions. Consider the energy change when an anhydrous salt gains its waters of hydration, or the reverse, the energy required to remove the water.

In the first case, there will be a lot of heat loss due to the rapidity of the reaction (steam). In the latter, we can't measure the temperature change when heating strongly.

The solution is to use Hess' Law and heats of solvation to determine the heat of hydration.

In this case we will wish to know the energy change when anhydrous magnesium sulfate gains its seven water molecules

$MgSO_4(s) + 7H_2O(l) \rightarrow MgSO_4 \cdot 7H_2O(s)$

But it's more convenient to react the reactant and the product separately with water

$MgSO_4(s) + 100H_2O(l) \rightarrow MgSO_4(aq, 100H_2O)$ ΔH_1 (heat of solvation of $MgSO_4$)

$MgSO_4 \cdot 7H_2O(s) + 93H_2O(l) \rightarrow MgSO_4(aq, 100H_2O)$ ΔH_2 (ΔH_{solv} of $MgSO_4 \cdot 7H_2O$)

By reversing the second equation, we can then cancel out the solutions, and 93 waters

$MgSO_4(s) + \cancel{100}\ 7H_2O(l) \rightarrow \cancel{MgSO_4(aq, 100H2O)}$ ΔH_1

$\cancel{MgSO_4(aq, 100H_2O)} \rightarrow MgSO_4 \cdot 7H_2O(s) + \cancel{93H_2O(l)}$ $-\Delta H_2$

$MgSO_4(s) + 7H_2O(l) \rightarrow MgSO_4 \cdot 7H_2O(s)$ $\Delta H_1 - \Delta H_2 = \Delta H_{hydration}$

Reaction of a solid and a solution

When we consider the reaction of a solid with a dissolved substance, we have to consider the dissolving of the solid and then the reaction of the two solutions.

For example, we can react solid NaOH with hydrochloric acid.

1. $NaOH(s) + HCl(aq) \rightarrow NaCl(aq) + H_2O(l)$
2. $NaOH(s) + H_2O(l) \rightarrow NaOH(aq)$
3. $NaOH(aq) + HCl(aq) \rightarrow NaCl(aq)$

If we add equations 2 and 3 we will obtain the same result as equation 1.

Bond Enthalpy

Bond enthalpy is the energy required to break one mole of a certain type of bond in the gaseous state averaged across a variety of compounds.

This means that for example there are lots of compounds with a C-H bond, so many different compounds are used to determine the C-H bond strength, and the average is then calculated. So, no single compound has the same value of C-H bond strength. It is often slightly different from the experimentally determined values.

Bond enthalpies are only an approximation, as they are not specific to any individual compound. However there are a few instances where they are "exact" - these occur with the diatomic elements eg; H_2, Cl_2, O_2 etc. and other diatomic compounds eg; HCl, HF, etc.

Take note

There are values given in the data book for bond enthalpy, but how do you use them? What's the formula? Consider Figure 5.5. We want to move from reactants to products (light blue), but have to go via atoms. We need to break the reactants (red arrow), and then make the products (green).

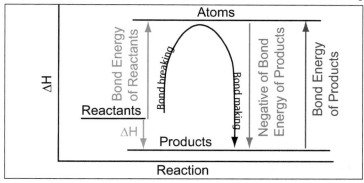

Figure 5.5: Bond Enthalpy Potential Energy diagram

Because we are taking the opposite value of the bond enthalpy of the products the mathematical formula becomes...

$$\Delta H = \Sigma \text{ B.E.}_{(reactants)} - \Sigma \text{ B.E.}_{(products)}$$

Let's use the Haber process as a classic example.

Example

$$N_2(g) + 3H_2(g) \rightleftharpoons 2NH_3(g)$$

You have to draw the molecules out to get a better feel for what you are dealing with. Draw the Lewis structures in the margin.

Strength of the reactant bonds	Strength of the product bonds
(N≡N) + 3(H-H)	6(N-H)
(944 kJ) + 3(436 kJ)	6(388 kJ)
2252 kJ	2328 kJ

So it takes more energy to break the products. From this we know that the products are more stable, and therefore should be at a lower potential energy (enthalpy) than the products. This means the reaction is exothermic and the value should be negative.

$\Delta H = \Sigma \text{ B.E.}_{(reactants)} - \Sigma \text{ B.E.}_{(products)}$

$\Delta H = 2252 \text{ kJ} - 2328 \text{ kJ}$

$\Delta H = -74 \text{ kJ}$ (exothermic as predicted)

Let's try it as a Hess' Law group of equations.

$N_2(g) \rightarrow 2N(g)$	944 kJ
$3H_2(g) \rightarrow 6H(g)$	3(436 kJ)
$2NH_3(g) \rightarrow 2N(g) + 6H(g)$	6(388 kJ)

Clearly we have to reverse (flip) the last equation so that the product can be made from the gaseous atoms. This will lead to the following:

$N_2(g) \rightarrow$ ~~$2N(g)$~~	944 kJ
$3H_2(g) \rightarrow$ ~~$6H(g)$~~	3(436 kJ)
~~$2N(g) + 6H(g)$~~ $\rightarrow 2NH_3(g)$	-6(388 kJ)

Cancellation gives us our desired equation and enthalpy of reaction.

$N_2(g) + 3H_2(g) \rightleftharpoons 2NH_3(g)$ $\Delta H = -74$ kJ

5.2 Learning Check

Determine the enthalpy of reaction for the following reactions. Be sure to draw out the correct Lewis structure of the molecules so that you identify the number and type of bonds. Use your data book.

1. $CH_4 + 2O_2(g) \rightarrow CO_2 + 2H_2O$
2. $H_2C=CH_2(g) + Br_2 \rightarrow H_2BrC-CBrH_2$
3. $H_2N-NH_2 + O_2 \rightarrow N_2 + 2H_2O$
4. $H_2 + Cl_2 \rightarrow 2HCl$
5. $CH_3COOH + 2O_2 \rightarrow 2CO_2 + 2H_2O$

Standard Enthalpy Changes

Definition

Standard Enthalpy changes are those changes which occur when a reaction proceeds under standard ambient temperature and pressure (SATP) conditions.

Standard Conditions – 25°C / 298K; 1 atm / 100.0 kPa, concentration = 1.0 M
These conditions are noted with a ° symbol - eg $\Delta H°$

Don't confuse these conditions with the standard state for a gas (0°C) - seems there's a double standard!

Standard State – the physical state (solid, liquid, gas) of a substance at standard conditions.

Standard Enthalpy of Formation

Standard Enthalpy (Heat) Change of Formation $\Delta H°_f$ – The energy change associated with the formation of ***one mole*** of substance from its ***elements*** in their ***standard states.***

NB: * the $\Delta H°f$ for an element in its standard state is defined as zero.

The Logic: In the standard heat of formation, we imagine forming both reactants and products from their elements in standard states. However, in order to move from reactants to products via the elements, we need to do the opposite process for the reactants. We need to find the value for the enthalpy of reaction (light blue), but we need to follow the path from the reactants to the elements (green), and then down to the products (dark blue).

Note

Compound	$\Delta H°_f$
$Al_2O_3(s)$	-1676.0
$CH_3Cl(g)$	-82.0
$CH_4(g)$	-74.9
$CO_2(g)$	-394.0
$CO(NH_2)_2(s)$	-333.0
$Fe_2O_3(s)$	-822.2
$H_2O(l)$	-286.0
$H_2O_2(l)$	-187.8
$HCl(g)$	-92.5
$NaCl(s)$	-413.0
$NaOH(s)$	-426.8
$NH_3(g)$	-46.0

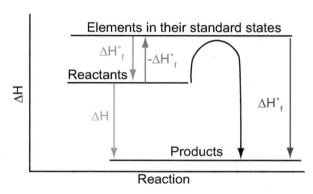

Figure 5.6: Standard Heat of Formation Potential Energy diagram

$$\Delta H = \Sigma \Delta H°_{f(products)} - \Sigma \Delta H°_{f(reactants)}$$

Ammonia can be combusted to produce NO_2 and water.

$4NH_3(g) + 7O_2(g) \rightarrow 4NO_2(g) + 6H_2O(g)$

Example

Determine the enthalpy of reaction given that $\Delta H°_f$ values are as follows...
$NH_3(g)$: -46 kJ mol⁻¹; $NO_2(g)$: 34 kJmol⁻¹; $H_2O(g)$: -286 kJmol⁻¹

Solution $\Delta H = \Sigma \Delta H°_{f(products)} - \Sigma \Delta H°_{f(reactants)}$
$\Delta H = [4NO_2(g) + 6H_2O(g)] - [4NH_3(g)]$
$\Delta H = [4(34 \text{ kJmol}^{-1}) + 6(-286 \text{ kJmol}^{-1})] - [4(-46 \text{ kJmol}^{-1})]$
$\Delta H = [136 \text{ kJmol}^{-1} + (-1716 \text{ kJmol}^{-1})] - [-184 \text{ kJmol}^{-1}]$
$\Delta H = [-1580 \text{ kJmol}^{-1})] - [-184 \text{ kJmol}^{-1}]$
$\Delta H = -1396 \text{ kJmol}^{-1}$

Use the values in the margin to calculate the enthalpy of reaction for the following.
1. $2H_2O_2(l) \rightarrow 2H_2O(l) + O_2(g)$
2. $HCl(g) + NaOH(s) \rightarrow NaCl(s) + H_2O(l)$
3. $CH_4(g) + Cl_2(g) \rightarrow CH_3Cl(g) + HCl(g)$
4. $2NH_3(g) + CO_2(g) \rightarrow CO(NH_2)_2(s) + H_2O(l)$
5. $2Al(s) + Fe_2O_3(s) \rightarrow Al_2O_3(s) + 2Fe(s)$

5.3 Learning Check

Standard Enthalpy of Combustion

Standard Enthalpy Change (Heat) of Combustion $\Delta H°_c$ – the energy change when one mole of the substance in its standard state is burned in excess oxygen.

The Logic: In order to move from reactants to products via the combustion products, often CO_2 and H_2O, we need to use the opposite process for the products. Our answer is ΔH (light blue), but we need to follow the combustion of the reactants (red) and then back up to the products (green).

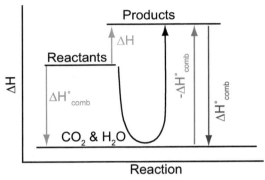

Figure 5.7: Standard Heat of Combustion Potential Energy diagram

$$\Delta H = \Sigma \Delta H°_{comb(reactants)} - \Sigma \Delta H°_{comb(products)}$$

Example

Use values from your data book to determine the enthalpy of reaction for the hydrogenation of ethene, $C_2H_4(l) + H_2(g) \rightarrow C_2H_6(g)$

$\Delta H = \Sigma \Delta H°_{comb}(reactants) - \Sigma \Delta H°_{comb}(products)$
$\Delta H = \Sigma [(C_2H_4) + (H_2)] - [C_2H_6]$
$\Delta H = \Sigma [(-1409 \text{ kJ mol}^{-1}) + (-286 \text{ kJ mol}^{-1})] - (-1560 \text{ kJ mol}^{-1})$
$\Delta H = [-1695 \text{ kJ mol}^{-1}] - (-1560 \text{ kJ mol}^{-1})$
$\Delta H = -105 \text{ kJ mol}^{-1}$

The hydrogenation of ethene is exothermic; $\Delta H = -105 \text{ kJ mol}^{-1}$

5.4 Learning Check

Determine the enthalpy of reaction for the following reactions using combustion data from the data book.

a) $C_6H_6(l) + 3H_2(g) \rightarrow C_6H_{12}$
b) $C_8H_{18} \rightarrow 2C_2H_4 + C_4H_{10}$

Which formula do I use?

How do you remember what to subtract from what. The secret is in the definitions.

Bond Enthalpy is about **breaking reactant** bonds (and making products).
Heat of formation is about **forming products**.
Heat of combustion is about **burning reactants**.

Quantity	Definition Focus	Calculation
Bond Enthalpy	reactants	reactants - products
Heat of Formation	products	products - reactants
Heat of Combustion	reactants	reactants - products

Energetics

Born – Haber Cycles & Ionic Lattice Enthalpy

Born – Haber cycles are just another application of Hess' Law. The cycle demonstrates two pathways to get from elements in their standard states to the compound.

The shortest pathway is the Standard Heat of Formation ($\Delta H°_f$)

The longest pathway is step – by – step from the element to the ions.

Consider lithium fluoride. Solid lithium must be turned into a positive gaseous ion and the fluorine molecule must be turned into a negative gaseous ion. Then the two ions must come together to form the solid lattice.

Because it is a cycle, all the values must add up to zero – you get back to where you started.

$$\Delta H_{atom\,Li} + \Delta H_{I.E.} + \Delta H_{atom\,F} + \Delta H_{E.A.} + \Delta H_{lattice} - \Delta H°_f = 0$$

Lattice Enthalpy – a special note – different authorities define it differently – for some it is the energy required to separate the ionic lattice into an ionic gas (endothermic), and for others it is the energy released when the separate gaseous ions "collapse" into the lattice (exothermic). There is no difference in the absolute value, but the sign will change accordingly, so make sure you know what the question is asking for.

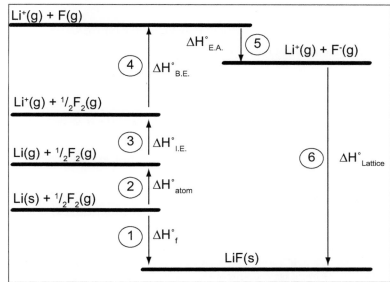

Figure 5.8: Born-Haber Cycle for lithium fluoride

1	$Li(s) + ½F_2(g) \rightarrow LiF(s)$	Heat of Formation ($\Delta H°_f$)	formation of 1 mole of the compound (usually exo)
2	$Li(s) \rightarrow Li(g)$	Atomisation (ΔH_{atom})	Lithium solid must be turned into a gas (endo)
3	$Li(g) \rightarrow Li^+(g) + e^-$	First ionization Energy ($\Delta H_{I.E.}$)	Lithium gas must be ionized (endo)
4	$F_2(g) \rightarrow 2F(g)$	Atomisation (ΔH_{atom})	F–F bond must be broken (endo)
5	$F(g) + e^- \rightarrow F^-(g)$	Electron Affinity ($\Delta H_{E.A.}$)	F atoms gain an electron (exo)
6	$Li^+(g) + F^-(g) \rightarrow LiF(s)$	Lattice Enthalpy ($\Delta H_{lattice}$)	ions are attracted into a lattice (exo)

Table 5.9: Reactions involved in a Born-Haber Cycle

When solving a B-H cycle, don't forget that Heat of Formation is pointing the wrong way and will need to be subtracted (sign reversed). Draw a clock-wise circle in Figure 5.8 and note that Step 1 is the only reaction that needs to be reversed.

The principle factor affecting lattice enthalpy is the charge density of the ions.

Compound	Melting point °C	comment
NaCl	801	both single charged ions - our reference point
Na_2O	1132	oxygen smaller and double negative charge
$MgCl_2$	1412	magnesium smaller and double positive charge
MgO	2600	both ions doubly charged.

Entropy

Entropy is a measure of how many ways the available energy may be distributed among the particles. For example, rotational motion, bond vibrations (stretching & bending). If a reaction produces more particles, then there are more places to "put" the energy.

Because gases are free to move in any direction, they have high entropy.

Everything has a certain amount of entropy. Zero entropy is defined as a "perfect" crystal at absolute zero.

Entropy is the ultimate decider if a reaction is to be spontaneous or not. But be careful. There are two ways entropy affects the universe.

Entropy of the *surroundings* is affected by the enthalpy change. If a process is exothermic, then the particles of the surroundings will gain kinetic energy.

Entropy of the *system* is affected by the chemical reaction or process occurring. .

There are four ways to increase the entropy of the system.

1. Mixing	mixtures have more ways to distribute the energy compared to separate pure substances
2. Increase of the number of particles	a reaction which yields an increase in the number of particles will increase the entropy.
3. Change of State	solids have the least entropy, liquids more, and gases the most.
4. Creation of a gas	reactions that produce gas will increase the entropy more than the other three factors.

Table 5.10: Entropy Changes of the system

5.5 Learning Check

Identify the sign of the entropy changes in the following processes

1. $H_2O(l) \rightarrow H_2O(g)$
2. $NaCl(s) + H_2O(l) \rightarrow NaCl(aq)$
3. $NH_3(g) + HCl(g) \rightarrow NH_4Cl(s)$
4. $2C_2H_6(l) + 7O_2(g) \rightarrow 4CO_2(g) + 6H_2O(g)$
5. $2HI(g) \rightarrow H_2(g) + I_2(g)$
6. $2HI(g) \rightarrow H_2(g) + I_2(s)$

Energetics

Hess' Law and Entropy

We can use Hess' Law for any of the state functions. Because entropy is always associated with the formation of a substance it will follow the formula

$$\Delta S = \text{sum}(S_{products} - S_{reactants})$$

Spontaneity & Gibb's Free Energy

Spontaneity is the result of the two forces that drive reactions – enthalpy and entropy.

Spontaneity is how we determine whether or not a reaction is going to proceed or not.

There are two driving forces in nature regarding chemical reactions – the drive to lower potential energy (exothermic reactions) and the drive to maximum randomness (an increase in entropy). If both of these driving forces are satisfied in a chemical reaction then the reaction will occur without any outside influences. If neither of these driving forces are satisfied, then the reaction will never occur unless work is done. If one or the other of these driving forces is satisfied, then the surrounding temperature will determine whether or not the reaction will occur.

Consider the evaporation and condensation of water.

Evaporation leads to an increase in randomness – entropy increase the entropy of the surroundings. Condensation leads to a decrease in potential energy – enthalpy decrease the enthalpy of the system.

Which process occurs? – It depends upon the temperature. Above a certain temperature the evaporation process is favoured. Below a certain temperature the condensation process is favoured. We know this special temperature as the boiling point.

$$\Delta G = \Delta H - T\Delta S$$

Formula

In considering whether ΔG will be positive or negative, we need to consider the balance of the two terms in the equation, ΔH and $T\Delta S$.

There are four possible scenarios to consider. These are best illustrated below.

		Enthalpy	
		Exothermic $\Delta H < 0$	Endothermic $\Delta H > 0$
Entropy	$\Delta S > 0$	(1) $\Delta G = \Delta H - T\Delta S$ $\Delta G = (-) - T(+)$ A negative minus a positive is always negative, therefore the reaction is spontaneous at all temperatures.	(2) $\Delta G = \Delta H - T\Delta S$ $\Delta G = (+) - T(+)$ A positive minus a positive is only negative if the $T\Delta S$ term is larger, therefore the reaction is spontaneous at high temperatures. (This is cooking)
	$\Delta S < 0$	(3) $\Delta G = \Delta H - T\Delta S$ $\Delta G = (-) - T(-)$ a negative minus a negative is only negative if the $T\Delta S$ term is small, therefore the reaction is spontaneous at low temperatures	(4) $\Delta G = \Delta H - T\Delta S$ $\Delta G = (+) - T(-)$ a positive minus a negative will always be positive, therefore the reaction is never spontaneous.

Situation (1) – the reaction is heating up the surroundings, which increases the entropy of the surroundings, and the system is also becoming more random – it's a win/win situation for spontaneity.

Situation (2) – The system is gaining potential energy, but at the expense of the entropy of the surroundings. Temperature matters! Think about it - you have to heat an endothermic reaction!

Situation (3) – The system is moving towards a lower potential energy at the expense of the increase in order of the surroundings. Temperature matters - exothermic processes are more spontaneous when it is cold.

Situation (4) – No one wins – the enthalpy is increasing, and the entropy is decreasing. Neither of the driving forces are being satisfied. Temperature is not a factor.

Generally Examiners are interested in situations 2 & 3 because temperature is a factor.

Watch out for units!!!

Exam Trap

ΔH is often quoted in **kilo**joules, while ΔS is in Joules per Kelvin.

Temperature is often given in Celsius - convert to Kelvin.

Free Energy & Equilibrium

In between the spontaneous reaction, where ΔG is negative, and the non-spontaneous reaction where ΔG is positive, we have ΔG=zero. This is an important situation because it represents the balance between the spontaneous and non-spontaneous reaction. This will represent the equilibrium situation or the temperature at which the reaction just occurs..

Example

Let's look at the following example....

Consider the endothermic reaction $NH_4Cl(s) \rightarrow NH_3(g) + HCl(g)$

a) The value of the enthalpy change is 176 kJ mol^{-1}. Determine the sign of the enthalpy change.
b) The value of the entropy change is 284 J mol^{-1} K^{-1}. Determine the sign of the entropy change.
c) Is this reaction spontaneous under standard conditions?
d) What effect will increasing temperature have on the spontaneity?
e) Determine the temperature, in degrees Celsius, at which the decomposition is spontaneous.

Solution

a) Endothermic reaction means positive ΔH.
b) Two moles of gas produced, therefore entropy is increasing.
c) $\Delta G = \Delta H - T\Delta S$
   $\Delta G = +176000$ J mol^{-1} — (298K · 284 J mol^{-1}K^{-1})
   $\Delta G = + 91368$ J mol^{-1}
   $\Delta G = + 91.4$ kJ mol^{-1}
d) Increasing the temperature will increase the spontenaity because as the temperature increases the $T\Delta S$ term increases and will become larger than the ΔH term.
e) At the point when the two terms are equal, $\Delta G = 0$.
   $0 = \Delta H - T\Delta S$
   $T = \Delta H \div \Delta S$
   $T = 176000$ kJ mol-1 $\div$ 284 kJ mol-1K-1
   $T = 620$ K
   $T = 620 - 273 = 346°C$

Energetics

Summary Questions

1. Determine the heat evolved in the reaction...
$$Al_2O_3(s) + 6Na(s) \rightarrow 2Al(s) + 3Na_2O(s)$$

   Given
   $$2Al(s) + {}^3/_2O_2(g) \rightarrow Al_2O_3(s) \quad \Delta H = -1590 \text{ kJ mol}^{-1}$$
   $$2Na(s) + {}^1/_2O_2(g) \rightarrow Na_2O(s) \quad \Delta H = -422.6 \text{ kJ mol}^{-1}$$

2. Determine the $\Delta H_{lattice}$ for KCl given the following information.
   $\Delta H°_f = -436$ kJ mol^{-1}
   $\Delta H_{atom}(K) = +90$ kJ mol^{-1}
   $\Delta H_{atom}(Cl_2) = +121$ kJ mol^{-1}
   $\Delta H_{E.A.} = -364$ kJ mol^{-1}
   $\Delta H_{I.E.} = +420$ kJ mol^{-1}

3. When 250 cm^3 of 0.500 M KOH(aq) and 250cm^3 of 0.500 HCl(aq) are mixed in a insulated styrofoam container, the temperature rises from 24.6°C to 28.0°C. Determine the enthalpy of reaction.

4. How will the temperature change in Q3 change when...
   a) both volumes are doubled?
   b) both concentrations are doubled?
   c) the concentration of acid is doubled?
   d) the volume of acid is doubled?

5. Consider the following phase change...
$$Br_2(l) \rightleftharpoons Br_2(g)$$

   a) The value of ΔH is 31 kJ mol^{-1}, determine the sign.
   b) The value of ΔS is 93 J mol^{-1}K^{-1}, determine the sign.
   c) How will increasing the temperature affect the spontaneity of this process?
   d) Determine the boiling point of bromine, in Celsius.

C.Lumsden · IB HL Chemistry

Chapter 6

Kinetics

In this chapter...

- 100 Rates of Reaction
- 100 Rate Experiments
- 101 Collision Theory
- 101 Factors affecting Rate of Reaction
- 104 Rate Expression / Rate Law
- 107 The Rate Constant
- 108 Reaction Mechanisms
- 109 Activation Energy
- 110 Summary Questions

Rates of Reaction

Definition

$$\text{Rate of reaction} = \frac{\Delta[\text{Reactants}]}{\Delta \text{time}} = \frac{\Delta[\text{Products}]}{\Delta \text{time}}$$

Or in words Rate of reaction is the change in concentration over time.

Rates of reaction are "officially" measured in the change of concentration per second ($mol \cdot dm^{-3} \cdot s^{-1}$), but there are other units that are used practically.

In experiments, we don't have a way of measuring concentration directly so we have to convert from other measurements.

Rate Experiments

- Rate of gas production – collection of gas (with a syringe).
- Rate of gas production – loss of mass of reactants.
- Rate of change of concentration – absorption of light.
- Rate of increase in precipitate formation (cloudiness of reaction mixture) – blocking of light.
- Time to consume a reactant (simple) – e.g. time for Mg strip to react and dissolve in acid.
- Time to consume a reactant (clock reaction) – I_2 clock.

In all the above types of measurements, we must use the limiting reactant to determine the rate.

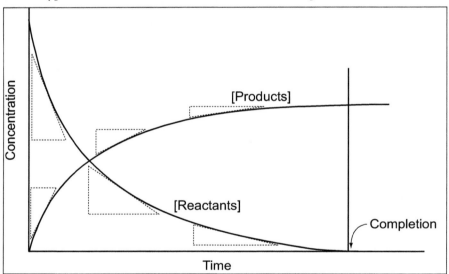

Figure 6.1: Reaction going to completion

As rate is the change of concentration over time, this is seen as the slope of the tangents in Figures 6.1 and 6.2.

If the graph of the reactants goes to zero the reaction has gone to completion. If the graph of the reactants levels out at some positive value, (i.e. there are still reactants remaining) then the reaction has reached equilibrium - Figure 6.2 - covered in the next chapter.

As you can see, the rate changes over time, so chemists of speak of the "initial rate", before reactant concentration can change.

Kinetics

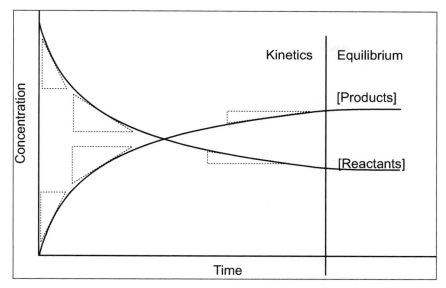

Figure 6.2: Reaction achieving equilibrium

Collision Theory

In order for a chemical reaction to occur the particles must...

- Collide,
- in the right orientation,
- with sufficient energy (more than the activation energy barrier).

Therefore the rate of reaction depends on the frequency of collisions, the geometry of the collisions and the number of particles that have more than the required activation energy.

Factors affecting Rate of Reaction

There are four factors which affect the rate of reaction. Each needs to be explained in terms of one or more of the points in the collision theory.

1. Concentration / Pressure
2. Surface Area
3. Temperature
4. Catalyst

Lets take each one in turn.

1. Concentration / Pressure.

The concentration of a solution or the pressure of a gas is simply the number of particles per unit volume. In a concentrated solution or a gas under high pressure, the particles are packed closer together so the number of collisions per unit time will increase.

2. Surface Area

Collisions can only occur on the surface of a solid. If you cut a potato in half, then the boiling water can reach the "inside" surface that it couldn't previously. So again the number of collisions per unit time will increase.

3. Temperature

As noted earlier, temperature is a measure of the kinetic energy of the particles. If the particles have more kinetic energy, then there will be an increase in the frequency of collisions. But that's not the whole explanation.

Because the particles have more energy, there are a greater number of particles with sufficient energy to overcome the energy barrier represented by the Activation Energy.

Activation Energy is the minimum amount of energy required for a chemical reaction to occur

This is demonstrated by two temperature curves of the "Maxwell-Boltzman Energy Distribution"

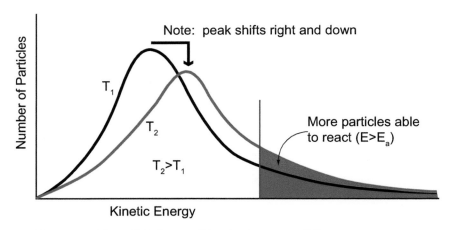

Figure 6.3: Maxwell-Boltzman Distribution at two different temperatures

In Figure 6.3, the blue area represents the amount (moles) of particles that have energy greater than the activation energy, and therefore are able to react at the lower temperature.

When the sample is heated to a higher temperature, the average energy of the particles increases, and the curve shifts to the right. The red area represents the increased number of particles that have sufficient energy to react at the higher temperature.

The entire area under one of the curves represents the total number of particles in the sample.

Common Mistake

You do not "give the particles activation energy". Activation energy is the energy barrier. You don't give runners "hurdles", you give them the energy to jump over the hurdles.

Note that the peak of the curve shifts to the right and downwards.

Exam Hint

You need to be able to draw and explain this graph if asked about temperature effects.

Kinetics

4. Catalyst.

A catalyst speeds up a chemical reaction by providing an alternate pathway (mechanism) that has a lower activation energy.

Catalysts are not consumed in a chemical reaction.

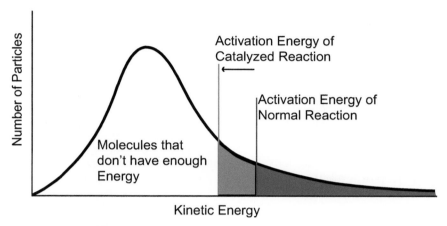

Figure 6.4: Maxwell Boltzmann distribution for a normal and a catalysed reaction

As in the definition, a catalyst lowers the activation energy so therefore there are a greater number of particles able to react at a given temperature.

In the case of both catalysts and increased temperature, there are more particles that have enough energy to react.

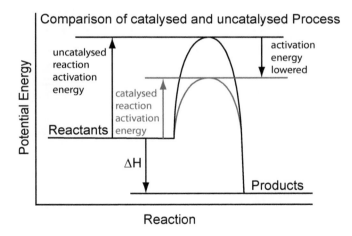

Figure 6.5: Energetic relationship for catalysed and uncatalysed exothermic reaction

While Figure 6.5 provides some evidence for the effect of a catalyst, Figure 6.4 is better as it clearly shows the increased number of particles with sufficient energy to react. Remember - when you have to talk about rates, you are talking about the number of particles per unit time that can react. The lowering of the activation energy is only the first part of the explanation.

Rate Expression / Rate Law

The determination of the Rate Expression is a common question to start off Paper 2.

The Rate Expression is **always experimentally determined**. There is no relationship to the balanced equation necessarily.

The rate expression is of the form Rate=k[A]^m[B]^n...
---
do not say "R=k[A]..."

      Where k is the rate constant,

      [A] and [B] are the reactants' concentrations,

      m is the order with respect to A,

      n is the order with respect to B,

      m+n is the overall order of reaction.

Looking at the graphical representations for each situation, we can see how the concentration of a reactant affects the rate of reaction.

Zero Order	First Order	Second Order
No Effect	Proportional	Exponential
Rate vs [Reactant] (flat line)	Rate vs [Reactant] (linear increase)	Rate vs [Reactant] (exponential curve)
The reactant has no effect on the rate. (anything to the power zero is one - multiplying by one has no effect.	The reactant has a linear effect on the rate. (double [A] = double rate)	The reactant has a exponential effect on the rate (double [A] = quadruple rate)

Table 6.6: Graphical Representations of Order of Reaction

Kinetics

Let's take the following four simple examples of rate

Example 1

For the reaction

$$A + B \rightarrow C$$

Trial	[A] / mol·dm^{-3}	[B] / mol·dm^{-3}	Rate of formation of C /mol·dm^{-3}·s^{-1}
1	1	1	2
2	1	2	4
3	2	2	16

If we compare trials 1 & 2, we can see that [A] is constant, [B] is doubling and the Rate is also doubling. Therefore [B] has a direct effect on the rate, when one doubles, the other doubles. That's first order (linear), so the order with respect to [B] is one.

Comparing experiments 2 & 3, we can see that [B] is constant, and [A] is doubling. The doubling of [A] causes a quadruple or four fold increase in the rate. $4 = 2^2$, so therefore the order with respect to [A] is two.

Our rate law is therefore
Rate = k[A]^2[B]^1 or because anything to the power 1 is itself...
Rate = k[A]^2[B]
This is 3rd order overall.

Example 2

For the reaction $D + E \rightarrow F$

Expt #	[D] / mol·dm^{-3}	[E] / mol·dm^{-3}	Rate of formation of F /mol·dm^{-3}·s^{-1}
1	0.1	0.01	6
2	0.1	0.02	6
3	0.2	0.04	12

Looking at experiments 1 & 2, there is no change in rate, so the doubling of [E] has no effect. Therefore the order with respect to [E] is zero
As [D] is doubled, the rate also doubles, and since [E] has no effect we ignore it. Therefore, the reaction is first order with respect to [D]:

Rate = k[D]^1[E]^0 or because anything to the power of zero is one...
Rate = k[D]

For the reaction G + H → J

Expt #	[G] / mol·dm^{-3}	[H] / mol·dm^{-3}	Rate of formation of J /mol·dm^{-3}·s^{-1}
1	1	1	5
2	1	2	10
3	2	4	80

This is the tricky one, because you don't have [H] constant between any of the different experiments, therefore it is harder to determine the effect of [G]. But we do know how [H] affects the rate. As [H] doubles, the rate doubles. Therefore the rate is first order in [H].
Comparing experiments 2 & 3, both [G] and [H] are doubling. The rate is increasing by a factor of 8 (2x2x2), which is 2^3. We can assign one of those doublings to [H]. The doubling of [G] causes the remaining four-fold increase in the Rate. Therefore the order with respect to [G] is two. So the rate expression is:

Rate = k[G]^2[H]

Example 4

For the reaction L + M + N → 2P + Q

Expt #	[L] / mol·dm^{-3}	[M] / mol·dm^{-3}	[N] / mol·dm^{-3}	Rate of formation of Q /mol·dm^{-3}·s^{-1}
1	1	1	3	5
2	1	4	3	20
3	2	8	3	160
4	2	8	9	1440

Experiments 1 & 2 look at the effect of [M] as all other species are constant. As [M] increases by a factor of four, so does the rate, so therefore the order with respect to [M] is one.

Looking at experiments 3 & 4, only [N] is changing - by a factor of 3. What is the effect on the Rate?
1440 ÷ 160 = 9. So as the concentration of N triples, the rate increases by a factor of 9. Since 9=3^2, the reaction is second order with regards to [N] (Watch out for this one, as it catches sloppy maths.)

Lastly, what is the effect of [L]? We have to look at experiments 2 & 3. As [L] doubles, so does [M] and the rate increases by a factor of 160÷20 = 8. We already know that the order with respect to [M] is one, 8 = 2x2x2. One of those 2's is from the change in [M]. Therefore the remaining change is due to [K], and must be 4 = 2^2. Therefore the rate is second order with respect to [L]

Rate = k[L]^2[M][N]^2

The Rate Constant

You will almost always be asked to determine the value and units of the rate constant.
You can do this by simple substitution of values from one of the experiments and solve for k.
For clarity, the value and the units are done separately here - in each case for trial #1.

Example 1	Example 2	Example 3	Example 4
$k = \dfrac{\text{Rate}}{[A]^2[B]}$	$k = \dfrac{\text{Rate}}{[D]}$	$k = \dfrac{\text{Rate}}{[G]^2[H]}$	$k = \dfrac{\text{Rate}}{[L]^2[M][N]^2}$
$k = 2 \div \{(1)^2(1)\}$	$k = 6 \div (0.1)$	$k = 5 \div \{(1)^2(1)\}$	$k = 5 \div \{(1)^2(1)(3)^2\}$
$k = 2$	$k = 60$	$k = 5$	$k = 0.56$

The units...

Example 1	Example 2
$k = \dfrac{mol \cdot dm^{-3} s^{-1}}{(mol \cdot dm^{-3})^2 (mol \cdot dm^{-3})}$ $k = \dfrac{s^{-1}}{(mol \cdot dm^{-3})^2}$ $k = dm^6 \cdot mol^{-2} \cdot s^{-1}$	$k = \dfrac{mol \cdot dm^{-3} s^{-1}}{(mol \cdot dm^{-3})}$ $k = s^{-1}$

Example 3	Example 4
$k = \dfrac{mol \cdot dm^{-3} s^{-1}}{(mol \cdot dm^{-3})(mol \cdot dm^{-3})}$ $k = \dfrac{s^{-1}}{(mol \cdot dm^{-3})^2}$ $k = dm^6 \cdot mol^{-2} \cdot s^{-1}$	$k = \dfrac{mol \cdot dm^{-3} s^{-1}}{(mol \cdot dm^{-3})^2 (mol \cdot dm^{-3})(mol \cdot dm^{-3})^2}$ $k = \dfrac{s^{-1}}{(mol \cdot dm^{-3})^4}$ $k = dm^{12} \cdot mol^{-4} \cdot s^{-1}$

Example 1: Rate = 2 $dm^6 \cdot mol^{-2} \cdot s^{-1}$ $[A]^2[B]$

Example 2: Rate = 60 s^{-1} $[D]$

Example 3: Rate = 5 $dm^6 \cdot mol^{-2} \cdot s^{-1}$ $[G]^2[H]$

Example 4: Rate = 0.56 $dm^{12} \cdot mol^{-4} \cdot s^{-1}$ $[L]^2[M][N]^2$

Overall order	Units of the rate constant
1	s^{-1}
2	$mol^{-1} \cdot dm^3 \cdot s^{-1}$
3	$mol^{-2} \cdot dm^6 \cdot s^{-1}$
4	$mol^{-3} \cdot dm^9 \cdot s^{-1}$
5	$mol^{-4} \cdot dm^{12} \cdot s^{-1}$

Table 6.7: The relationship between order and rate constant units

Reaction Mechanisms

A **reaction mechanism** is the sequence of elementary steps that make up the overall reaction.

An **elementary step** is a reaction whose rate law can be written from its molecularity.

Molecularity is the number of species colliding in an elementary step.

A molecularity of "1" is unimolecular (a unimolecular decomposition)
Molecularities of "2" are bimolecular collisions between two particles
Molecularities of "3" are termolecular and statisically unlikely.

Consider a car crash – It's very (very, very) unlikely that a 4 car pile up occurs with all four cars colliding simultaneously. More likely, two cars collide, then one more, perhaps knocking out one of the originals and then a fourth, perhaps combining with the second car etc.

A valid mechanism has two features. First it must agree with the balanced overall equation. Second the rate determining step (RDS) must agree with the experimentally determined rate law.

Rate Determining steps

The rate determining step is...

1) the elementary step with the highest activation energy and therefore...

2) is the slowest step. This is the step where you would want a catalyst to be involved. It's no use catalysing a fast step if your reaction rate is still limited by another step.

Example

Consider the reaction

$$A + 2B \rightarrow C + 2D$$

There are two possible mechanisms proposed...

John's proposed mechanism
Step 1: $A + B \rightarrow D + X$
Step 2: $B + X \rightarrow C + D$

Kate's proposed mechanism
Step 1: $B + B \rightarrow D + Y$
Step 2: $A + Y \rightarrow C + D$

If Step 1 is the slow step (RDS)

Mechanism A
Rate = $k[A][B]$

Mechanism B
Rate = $k[B]^2$

If step 2 is the slow step (RDS)

Mechanism A
Rate = $k[B][X]$
because X relies on [A][B]
Rate = $k[A][B][B]$
Rate = $k[A][B]^2$

Mechanism B
Rate = $k[A][Y]$
because [Y] relies on [B][B]
Rate = $k[A][B][B]$
Rate = $k[A][B]^2$

The experimentally determined rate law will agree with one of these rate expressions and we have a good idea of what the mechanism could be. In this case, if the second step is slow, we cannot identify a correct mechanism.

Activation Energy

Activation energy is related to the rate constant, k, by the Arrhenius equation (in the data book).

$$k = Ae^{\frac{-E_a}{RT}}$$

Where...

k is the rate constant

A is the "frequency factor" - depends upon shape of molecules.

R is the ideal gas constant

E_a is the activation energy

T is the absolute temperature (Kelvin!!!!)

so strictly speaking the rate expression should look like:

$$Rate = Ae^{\frac{-E_a}{RT}}[X]^m[Y]^n$$

In order to make this formula more useable, we can take the natural logarithim of both sides and rearrange to have the equation in the form of a straight line.

$$\ln k = \ln A - \frac{E_a}{RT}$$

$$\ln k = -\frac{E_a}{R} \cdot \frac{1}{T} + \ln A$$

$$y = m \cdot x + b$$

Therefore graphing $ln(k)$ vs $1/T$ we can generate a line with a slope of $-E_a/R$.

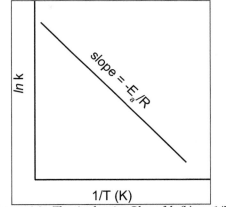

Figure 6.8: The Arrhenius Plot of $ln(k)$ vs. $1/T$

Practically in the lab, you would have to do a series of experiments of the same reaction at different temperatures to find the rate constant at each temperature.

Temperature effect on Rate Constant

Exam Hint

The value of the **rate constant** often approximately doubles with every 10°C increase in temperature; therefore the rate often approximately doubles with every 10°C increase in temperature. Usually, temperature has a greater effect on the rate than the concentrations.

Summary Questions

1. One of the versions of the iodine clock uses the reaction between peroxydisulphate ions and iodide ions.

$$S_2O_8^{2-}(aq) + 3I^-(aq) \rightarrow 2SO_4^{2-}(aq) + I_3^-(aq)$$

Rate data for the experiment was found.

$[S_2O_8^{2-}]$	$[I^-]$	Rate
0.040	0.034	5.6×10^{-4}
0.040	0.017	2.8×10^{-4}
0.080	0.017	5.6×10^{-4}

a) Determine the order with respect to $S_2O_8^{2-}$.
b) Determine the order with respect to I^-.
c) Write the Rate expression.
d) Determine the rate constant and units.

---

2. Consider the reaction

$H_2 + 2ICl \rightarrow I_2 + 2HCl$

There are two proposed mechanisms.

Mechanism A	Mechanism B
Step 1: $H_2 + ICl \rightarrow HI + HCl$	Step 1: $ICl + ICl \rightarrow I_2 + Cl_2$
Step 2: $HI + ICl \rightarrow HCl + I_2$	Step 2: $H_2 + Cl_2 \rightarrow 2HCl$

a) Predict the rate law if...
   i) Step 1 of Mechanism A is slow
   ii) Step 2 of Mechanism A is slow
   iii) Step 1 of Mechanism B is slow

b) Explain why a one step reaction is unlikely.

Chapter 7

Equilibrium

In this chapter...

- 112 Dynamic Equilibrium
- 112 Approaching equilibrium
- 112 Position of Equilibrium
- 113 LeChâtelier's Principle
- 114 Reaction Quotient
- 115 The meaning of K_c
- 116 Determining the Equilibrium Constant, K_{eq}
- 117 Determining Equilibrium Concentrations
- 119 Summary Questions

Dynamic Equilibrium

Equilibrium is related to rates of reaction.

Equilibrium is achieved when the **forward rate** of reaction and the **reverse rate** of reaction are **equal**.

Dynamic equilibrium exists in a closed system when there is no change in the macroscopic properties (i.e. concentrations).

Approaching equilibrium

Before two reactants are mixed, the concentration of the products is zero.

You should imagine that you can mix the reactants and not let them react - this will help you determine the initial concentrations.

After the two reactants mix, they react to form products, and the concentration of products increases.

As [product] increases, the rate of the reverse reaction must also increase (rate ∝ [product]). At some point in time, as more products are formed from the forward reaction, they have sufficient concentration to cause the reverse rate of reaction to be equal to the rate of the forward reaction.

It doesn't matter which direction you approach equilibrium from, at some point the two rates will equal.

Position of Equilibrium

As chemists, we are interested in the products of our reactions, not the reactants. So we want to know how the amount of products compares to the amount of reactants. This is the idea behind the equilibrium constant.

$$K_c = \frac{[\text{Products}]}{[\text{Reactants}]}$$

Lucky for us it is easy to write the equilibrium expression for any balanced equation. Consider the generic reaction

$$aA + bB \rightleftharpoons dD + eE$$

General Formula

$$\boxed{K_c = \frac{[D]^d[E]^e}{[A]^a[B]^b}}$$

(the concentration of each species is raised to the power of its coefficient)

Classic examples

Haber Process $\quad N_2(g) + 3H_2(g) \rightleftharpoons 2NH_3(g)$

$$K_c = \frac{[NH_3]^2}{[N_2][H_2]^3}$$

Contact Process $\quad 2SO_2(g) + O_2(g) \rightleftharpoons 2SO_3(g)$

$$K_c = \frac{[SO_3]^2}{[SO_2]^2[O_2]}$$

Equilibrium

Write the equilibrium expression for each of the following reactions.

1. $2H_2S(g) + 3O_2(g) \rightleftharpoons 2H_2O(g) + 2SO_2(g)$
2. $2N_2O_5(g) \rightleftharpoons 4NO_2(g) + O_2(g)$
3. $CO(g) + 2H_2(g) \rightleftharpoons CH_3OH(g)$
4. $4NH_3(g) + 3O_2(g) \rightleftharpoons 2N_2(g) + 6H_2O(g)$
5. $2N_2O(g) + 3O_2(g) \rightleftharpoons 4NO_2(g)$

7.1 Learning Check

LeChâtelier's Principle

When an equilibrium system is stressed, the system will shift the position equilibrium so as to remove the stress.

Many candidates lose points here, not because they don't understand, but because they are not thorough enough with their answers.

A classic example with degrees of answer quality…

"State and explain the effect of increasing pressure on the amount of product in the Haber Process, $N_2(g) + 3H_2(g) \rightleftharpoons 2NH_3(g)$"

Classic Question

1. It shifts right.
2. It increases the product.
3. It increases product by shifting right.
4. It increases the amount of NH_3 by shifting to the right.
5. It increases the amount of NH_3 by shifting the equilibrium to the right to remove the stress.
6. It increases the amount of NH_3 by shifting the equilibrium to the right to remove the stress because there are fewer moles of gas of products than of reactants.
7. An increase in pressure shifts the equilibrium to the right because as four moles of reactant gases become two moles of product gas, the pressure is returned to its original state.
8. There is an increase in the amount of _product_ because an increase in _pressure_ shifts the equilibrium to the right, as _four moles of gaseous reactant_ become _two moles of gaseous product_ the pressure is reduced.

Model Answer

C.Lumsden   IB HL Chemistry

Stress	Result
Increasing concentration	Shift away from the species being increased in order to lower the concentration.
Removal of a species	Shift towards that species in order to maintain its concentration. This is usually accomplished by adding something else that reacts with only that species. E.g. the addition of base will consume any H^+ ions, and vice versa. E.g. The addition of Ag^+(aq) ions will precipitate any halide ions, Cl^-(aq).
Increasing pressure by decreasing volume	Shift towards the side with fewer moles **of gas**.
Adding a catalyst	No effect as it increases the rate of the forward and reverse reaction equally.
Adding an inert gas (increase pressure)	No effect (because it does not increase the partial pressure of a reactant or product).
Increasing Temperature	**Affects the equilibrium constant** - see below. Favours the endothermic reaction.

Changes in pressure or concentration only force the equilibrium to shift left or right in order to restore the equality of the rates of reaction and maintain the equilibrium constant.

Be Careful

However, a change in temperature actually changes the value of K_c.

As reactions are non-symmetrical in terms of energy (one way is endo, the other exo), the temperature will have an effect.

K_c is dependant upon temperature. As temperature increases, the rate of the endothermic reaction direction increases. This will change the relative concentrations of reactants and products.

Often you are given information about the nature of the reactants and products and asked to answer the question based on that information.

Example

In the reaction $2NO_2(g) \rightleftharpoons N_2O_4(g)$, NO_2 is a orange/brown gas, and N_2O_4 is colourless.

State and explain the effect of increasing pressure on the colour of the gases in the container.

Answer
As the pressure is increased, the equilibrium will shift to the products (right) because there is only one mole of gas of product compared to two moles of reactant. Shifting to the right will lower the pressure. Therefore the colour will fade.

Reaction Quotient

So let imagine that we have the power to stop the chemicals from reacting. If we upset the equilibrium by putting in more reactants for example our ratio of products to reactants will be too low. The equilibrium "wants" to use these up, but what is the ratio before this occurs - this is the reaction quotient, Q.

The reaction quotient, **Q** is the ratio of products to reactants (with the associated exponents) at non-equilbirum concentrations.

We can use the Reaction Quotient, Q, to expand our understanding of the process of coming to equilibrium.

For the Haber Process, the equilibrium constant at 500°C is 6.0×10^{-2}. Determine the reaction quotient and therefore which of the following situations represents equilibrium. Which way will the equilibrium shift?

Equilibrium

	Concentrations of species		
	N_2	H_2	NH_3
a)	1.0×10^{-5}	2.0×10^{-3}	1.0×10^{-3}
b)	1.5×10^{-5}	3.54×10^{-1}	2.0×10^{-4}
c)	5.0	1.0×10^{-2}	1.0×10^{-4}

a)	b)	c)
$Q = \dfrac{[NH_3]^2}{[N_2][H_2]^3}$	$Q = \dfrac{[NH_3]^2}{[N_2][H_2]^3}$	$Q = \dfrac{[NH_3]^2}{[N_2][H_2]^3}$
$Q = \dfrac{(1 \times 10^{-3})^2}{(1 \times 10^{-5})(2.0 \times 10^{-3})^3}$	$Q = \dfrac{(2 \times 10^{-4})^2}{(1.5 \times 10^{-5})(3.54 \times 10^{-1})^3}$	$Q = \dfrac{(1 \times 10^{-4})^2}{(5.0)(1.0 \times 10^{-2})^3}$
$Q = 12500000$	$Q = 0.060$	$Q = 0.002$

In a) the reaction quotient is higher than the equilibrium constant, so the system is not at equilibrium, and further must shift to the left to lower the Q to Kc.

In b) the reaction quotient is the same as the equilibrium constant and therefore the system is at equilibrium.

In c) the reaction quotient is lower than the Kc, so the equilibrium must shift to the right to increase the value up to the value of Kc.

For the dissociation of ethanoic acid, $K_a = 1.8 \times 10^{-5}$

$$CH_3COOH \rightleftharpoons CH_3COO^- (aq) + H^+(aq)$$

a) $[CH_3COOH] = 1.0 \times 10^{-2}$; $[CH_3COO^- (aq)] = [H^+(aq)] = 1.0 \times 10^{-3}$
b) $[CH_3COOH] = 1.0 \times 10^{-2}$; $[CH_3COO^- (aq)] = [H^+(aq)] = 1.0 \times 10^{-6}$
c) $[CH_3COOH] = 2.5 \times 10^{-2}$; $[CH_3COO^- (aq)] = 2.0 \times 10^{-5}$; $[H^+(aq)] = 1.0 \times 10^{-3}$

7.2 Learning Check

The meaning of K_c

As chemists, we are interested in the products of our reactions, so in terms of K_c, we are interested in increasing the value of K_c, and thus increasing the extent of the reaction.

$K_c >>> 1$	Reaction goes **almost** to completion
$K_c > 1$	Equilibrium position favours products
$K_c \approx 1$	Reactant and product concentrations similar
$K_c < 1$	Equilibrium position favours reactants

Determining the Equilibrium Constant, K_{eq}

In order to determine equilibrium values, it's best to use an "ICE" chart - "Initial Concentration", "Change of Concentration", "Equilibrium Concentration".
The Initial concentrations are given, and the products are usually zero, but not necessarily - you will be told.
When filling out the "Change" line, you must remember that reactant concentrations are decreasing, therefore the change is negative for reactants and positive for products. Also, the change agrees with the stoichiometric ratios (balancing numbers).
The Equilibrium line is the result of the initial and change concentrations.

	Reactants	Products
Initial	Info given in question	usually zero
Change	negative	positive
Equilibrium	Initial – Change	Initial + Change

Example

1. Consider the following equilibrium

$$SO_2(g) + NO_2(g) \rightleftharpoons SO_3(g) + NO(g)$$

A 2.0 dm³ flask was filled with 4.0 mol of SO_2 and 4.0 mol of NO_2. At equilibrium it was found to contain 2.6 mol of NO.
Determine the value of K_c.

Solution: The given values are black, the red values are determined by the student. The change line was determined by the last column.

The initial concentrations in this case are 4.0 mol ÷ 2 dm³ = 2.0 M.

	SO_2	NO_2	SO_3	NO
Initial	2.0 M	2.0 M	0	0
Change	−1.3 M	−1.3 M	+1.3 M	+1.3 M
Equilibrium	0.7 M	0.7 M	1.3 M	1.3 M

$$K_c = \frac{[SO_3][NO]}{[SO_2][NO_2]}$$

$$K_c = \frac{(1.3)(1.3)}{(0.7)(0.7)}$$

$$K_c = 3.45$$

We are not usually concerned with the units of the constant. They depend on the exponents, and in this case they have cancelled out.

Example

2. A 4.0 dm³ flask initially contained 12.0 mol of SO_3. At equilibrium it contained 3.0 mol of SO_2. Determine the K_c for the equation

$$2SO_3(g) \rightleftharpoons 2SO_2 + O_2$$

	$2SO_3$	$2SO_2$	O_2
Initial	3.0 M	0 M	0 M
Change	−0.75 M	+0.75 M	0.75/2 = +0.375
Equilibrium	2.25 M	0.75 M	0.375 M

In the change line, the O_2 concentration is half of the SO_2 due to the coefficients in the balanced equation.

$$K_c = \frac{[SO_2]^2[O_2]}{[SO_3]^2}$$

$$K_c = \frac{(0.75)^2(0.375)}{(2.25)^2}$$

$$K_c = 4.17 \times 10^{-2}$$

Equilibrium

Determining Equilibrium Concentrations

You are also asked to use a known equilibrium constant value to determine the equilibrium concentration of the species.

While the maths here can become much more difficult (wait until University – unless you have a tough teacher), good news, the IB has put a limit on how hard this type of question can be.

The good news is that you can always make an assumption that allows you to avoid using the quadratic formula.

The condition that the IB puts on the question is that it is only for very small values of K_c.

Consider for a minute, what this means. If K_c is small, then there is only a small amount of products compared to reactants.

Any change in the reactant concentration will be negligible. If x is subtracted or added to a much larger number, you can ignore it.

---

3. The reaction $N_2O_4 \rightleftharpoons 2NO_2$ has a $K_{eq} = 4.0 \times 10^{-7}$.   *Example*

If the initial concentration of N_2O_4 is 0.1 M determine the concentration of the products at equilibrium.

	N_2O_4	$2NO_2$
Initial	0.1 M	0
Change	-x	+2x
Equilibrium	0.1 M - x	2x

In the following solutions, you will see that the 0.1 - x term is assumed to be equal to 0.1, due to the fact that x must be a very small number. We can only make this assumption when there is a sum or difference - you can't do this for the "2x" factor.

Quadratic	Small K_c assumption: $(0.1-x) \approx 0.1$
$K_c = \dfrac{[NO_2]^2}{[N_2O_4]}$	$K_c = \dfrac{[NO_2]^2}{[N_2O_4]}$
$K_c = \dfrac{(2x)^2}{0.1-x}$	$K_c = \dfrac{(2x)^2}{0.1-x}$
$4.0 \times 10^{-7} = \dfrac{4x^2}{0.1-x}$	$K_c = \dfrac{(2x)^2}{0.1}$
$4x^2 = 4.0 \times 10^{-7}(0.1-x)$	$4.0 \times 10^{-7} = \dfrac{4x^2}{0.1}$
$4x^2 + 4.0 \times 10^{-7}x - 4.0 \times 10^{-8} = 0$	$4x^2 = 4.0 \times 10^{-8}$
$x^2 + 1.0 \times 10^{-7}x - 1.0 \times 10^{-8} = 0$	$x^2 = 1.0 \times 10^{-8}$
$x = 1.000500125 \times 10^{-4}$	$x = 1.0 \times 10^{-4}$
$x = 1.0 \times 10^{-4}$	

The % difference between the two methods is only 0.05%, so the assumption was valid.

You can't always make this assumption, but you won't have to solve a quadratic - sometimes the maths can simplify without dropping the "x" term.

Remember that "x" is likely not your answer; refer back to your table to solve for the equilibrium concentrations.

Final Answer: $[NO_2] = 2x$
$[NO_2] = 2.0 \times 10^{-4}$ M

Example

4. The reaction of $2NOCl \rightleftharpoons 2NO + Cl_2$ has an equilibrium constant value of 1.6×10^{-5}.

   If 1.0 mol of NOCl is placed in a 2.0 dm³ container, determine the concentration of the products at equilibrium

	2NOCl	2NO	Cl₂
Initial	0.5 M	0	0
Change	-2x	+2x	x
Equilibrium	(0.5-2x) ≈ 0.5	2x	x

$$K_c = \frac{[NO]^2[Cl_2]}{[NOCl]^2}$$

$$K_c = \frac{(2x)^2(x)}{(0.5)^2}$$

$$1.6 \times 10^{-5} = \frac{4x^3}{(0.25)}$$

$$4.0 \times 10^{-6} = 4x^3$$

$$x^3 = 1 \times 10^{-6}$$

$$x = \sqrt[3]{1 \times 10^{-6}}$$

$$x = 1 \times 10^{-2}$$

Final Answer
$[NO] = 2x = 2.0 \times 10^{-2}$ M,
$[Cl_2] = x = 1.0 \times 10^{-2}$ M

7.3 Learning Check

1. Consider the decomposition of hydrogen iodide into its elements. If the concentration of HI, H_2 and I_2 are 2.2×10^{-2}, 1.0×10^{-3} and 2.5×10^{-2} respectively at equilibrium, what is the equilibrium constant?

2. At 773°C a closed system is found to contain 0.105 M of CO, 0.250 M H_2, and 0.00261 M CH_3OH at equilibrium. If the reaction equation is

$$CO(g) + 2H_2(g) \rightleftharpoons CH_3OH(g)$$

   Find the equilibrium constant.

3. NO_2 decomposes according to

$$3NO_2(g) \rightleftharpoons N_2O_5(g) + NO(g)$$

   If 0.20 mol of NO_2 are placed in a 4.0 dm³ vessel, and $K_c = 1.0 \times 10^{-11}$, calculate the equilibrium concentrations.

Spontaneity and Equilibrium

One of the understandings that we have had until now is that it is either the reactants (non-spontaneous) or the products (spontaneous - Figure 7.1) represent a position of minimum free energy.

If the reaction is to proceed to a position of minimum free energy (maximum entropy, minimum enthalpy) it may be that this position is not represented purely by the reactants or the products, but rather some point along the reaction pathway (See Figure 7.2) which has both reactants and products; i.e. an equilibrium position.

As the reaction proceeds towards the minimum value from an non-equilibrium position (represented by the reaction quotient, Q), the change in free energy(green arrows) approaches zero.

$\Delta G°$ represents the reaction that never happens - the reaction doesn't go to completion.

If this minimum position is part way along the way along to products, how can it be represented graphically?

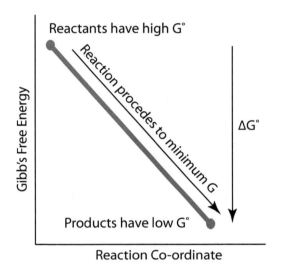

Figure 7.1: Spontaneous Reaction

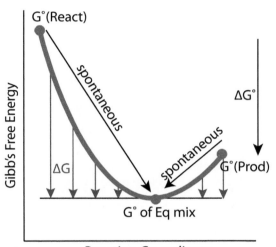

Figure 7.2: Equilibrium

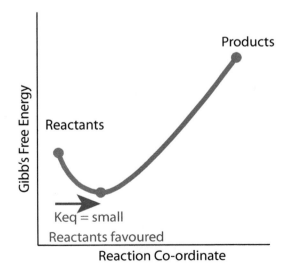

Figure 7.3: Equilibrium favours reactants

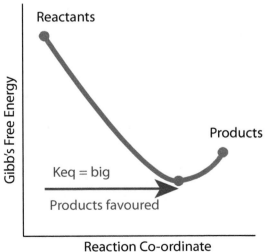

Figure 7.4: Equilibrium favours products

Equilibrium & ΔG° Calculations

As you would expect, there is a mathematical relationship between the Equilibrium Constant and Gibb's Free Energy values.

The relationship between G and K is given by $\Delta G = \Delta G° + RT\ln Q$. As ΔG approaches zero, Q approaches K_{eq}. $0 = \Delta G° + RT\ln K_{eq}$, which can be re-arranged to $\Delta G = -RT\ln K_{eq}$.

$\Delta G° = -RT\ln(K_{eq})$	$K_{eq} = e^{\frac{-\Delta G°}{RT}}$

Notice that we now have two ways to calculate ΔG. The first from Energetics, and the second from Equilibrium. We can combine these two equations to get the van't Hoff Equation

$$-RT\ln K = \Delta G = \Delta H - T\Delta S$$
$$-RT\ln K = \Delta H - T\Delta S$$
$$\ln K = \frac{\Delta H}{-RT} - \frac{T\Delta S}{-RT}$$
$$\ln K = \frac{-\Delta H}{R} \times \frac{1}{T} - \frac{\Delta S}{R}$$
$$y = mx + b$$

Which ultimately has the form of y=mx+b. So if we were to graph lnK vs. 1/T, we would get a straight line with a slope of $-\Delta H/R$. An exothermic reaction would have a positive slope and endothermic would have a negative slope.

One could also use this graph to determine ΔS, however due to the $\log_e$ scale on the y-axis, this will have high associated uncertainty.

Notice as well the very high similarity of the van't Hoff graph compared to the Arrhenius graph. Don't get them mixed up.

Example

What are the values of the equilibrium constant for the Haber Process (ΔG = -16.4 kJmol^{-1}) at 300K and 600K respectively?

$K_{eq} = e^{\frac{-\Delta G}{RT}}$	$K_{eq} = e^{\frac{-\Delta G}{RT}}$
$K_{eq} = e^{\frac{-(-16.4 kJmol^{-1})}{8.315 Jk^{-1}mol^{-1} \times 300K}}$	$K_{eq} = e^{\frac{-(-16.4 kJmol^{-1})}{8.315 Jk^{-1}mol^{-1} \times 600K}}$
$K_{eq} = e^{\frac{(16400 Jmol^{-1})}{8.315 Jk^{-1}mol^{-1} \times 300K}}$	$K_{eq} = e^{\frac{(16400 Jmol^{-1})}{8.315 Jk^{-1}mol^{-1} \times 600K}}$
$K_{eq} = e^{6.575}$	$K_{eq} = e^{3.288}$
$K_{eq} = 717$	$K_{eq} = 26.8$
Figure 7.5: Haber Equilibrium constant at 300K	Figure 7.6: Haber Equilibrium constant at 600K

Notice that this agrees with our expectation that the value for Keq decreases when the temperature conditions for an exothermic reaction increase.

Equilibrium

Summary Questions

1. Balance the following equations and write the equilibrium constant for the reactions.

   a) $N_2O_4(g) \rightleftharpoons NO_2(g)$
   b) $SiH_4(g) + Cl_2(g) \rightleftharpoons SiCl_4(g) + H_2(g)$
   c) $PBr_3(g) + Cl_2(g) \rightleftharpoons PCl_3(g) + Br_2(g)$
   d) $CO(g) + H_2(g) \rightleftharpoons CH_3OH(g)$
   e) $NO_2(g) \rightleftharpoons NO(g) + O_2(g)$

2. Nitric oxide reacts with oxygen exothermically to produce NO_2, which is a dark brown gas.

$$2NO(g) + O_2(g) \rightleftharpoons 2NO_2(g)$$

   How will the colour of the mixture change if...
   a) pressure is increased by reducing the volume of the container?
   b) temperature is increased?
   c) a catalyst is added?
   d) neon gas is added to increase the pressure?
   e) more oxygen is added?

3. At 25°C, 0.0560mol O_2 and 0.02 mol N_2O were placed in a 1.0 dm³ container and allowed to react according to the equation below.

$$2N_2O(g) + 3O_2(g) \rightleftharpoons 4NO_2(g)$$

   When the system reached equilibrium the concentration of the NO_2 was found to be 0.020 mol/dm³

   a) What were the equilibrium concentrations of N_2O and O_2?
   b) What is the value of K for this reaction at 25°C?

4. At 1000°C water decomposes according to the reaction below. If the equilibrium constant is 7.3×10^{-18} and the initial concentration of water is 0.100 M, find the equilibrium concentrations of hydrogen and oxygen.

   $2H_2O(g) \rightleftharpoons 2H_2(g) + O_2(g)$

Chapter 8

Acids & Bases

In this chapter...

- 122 Theories of Acids and Bases
- 122 Properties of Acids & Bases
- 123 $H^+(aq)$ or $H_3O^+(aq)$?
- 123 Conjugate pairs
- 124 Amphoteric substances
- 124 Strong & Weak
- 125 Distinguishing between Strong and Weak.
- 125 pH Scale
- 126 Acid Deposition
- 127 Calculation of pH of weak acid solutions
- 128 Hydrolysis of Salts
- 130 Buffer Solutions
- 131 Acid – Base Titration – Strong Acid & Strong Base
- 132 Acid – Base Titration – Weak & Strong
- 132 Indicators
- 133 pH Curves
- 134 Relative Strengths of Acids & Bases
- 135 Summary Questions.

Theories of Acids and Bases

Theory	Acid definition	Base definition
Brønsted – Lowry	H⁺(aq) donor	H⁺(aq) acceptor
Lewis	Electron pair acceptor	Electron pair donor

The Brønsted-Lowry theory is the "normal" definition that you probably learned before IB Chemistry. The Lewis definition is broader because it can describe :PCl$_3$ as a base and AlCl$_3$ as an acid.

Exam hint

Group III, electron deficient compounds are Lewis acids, and Group V compounds (with a lone pair) are Lewis bases.

Transition metal ions are Lewis acids, and the ligands around them are Lewis bases. Nucleophiles are Lewis bases and electrophiles are Lewis acids.

The result of these reaction is a co-ordinate bond.

Properties of Acids & Bases

There are five types of reactions that acids undergo. They are based on the five different types of base. In most cases, a base can be defined as anything that reacts with an acid. Bases often contain metals. The five reactions can be organized into three groups as follows...

Acids react with reactive metals to produce hydrogen gas and a salt.

2HCl(aq) + Mg(s) → MgCl$_2$(aq) + H$_2$(g)

Acids react with oxides and hydroxides to produce water and a salt

2HCl(aq) + MgO(s) → MgCl$_2$(aq) + H$_2$O(l)

2HCl(aq) + Mg(OH)$_2$(s) → MgCl$_2$(aq) + 2H$_2$O(l)

Acids react with carbonates and hydrogen carbonates to produce carbon dioxide, water and salt.

2HCl(aq) + MgCO$_3$(s) → MgCl$_2$(aq) + H$_2$O(l) + CO$_2$(g)

2HCl(aq) + Mg(HCO$_3$)$_2$(s) → MgCl$_2$(aq) + 2H$_2$O(l) + 2CO$_2$(g)

Alkalis are soluble bases. So all alkalis are bases, but not all bases are alkali.

8.1 Learning Check

Complete and balance the following equations

HNO$_3$(aq) + NaHCO$_3$(aq) →

Al$_2$O$_3$(s) + HCl(aq) →

ZnO(s) + H$_2$SO$_4$(aq) →

Mg(s) + HNO$_3$(aq) →

H$_2$SO$_4$(aq) + CuCO$_3$(s) →

HCl(aq) + Ca(OH)$_2$(aq) →

$H^+(aq)$ or $H_3O^+(aq)$?

Either notation is fine. Strictly speaking, $H_3O^+(aq)$ - the "hydronium ion" is more correct. This text will use $H^+(aq)$ for convenience, which is acceptable for you to write on exams.

Definition

It will save some confusion if you always react the acid or base with H_2O as follows

If we consider the acid "HA"

$$HA(aq) + H_2O(l) \dashrightarrow H_3O^+(aq) + A^-(aq)$$

or the base "B"

$$B(aq) + H_2O(l) \dashrightarrow HB^+(aq) + OH^-(aq)$$

It makes more sense to use the H_3O^+ to represent the acid, as there aren't really any bare protons floating around in the water - they protonate the water molecules.

Conjugate pairs

A conjugate pair are two species differing by a single proton, H^+

The notion of conjugate pairs only applies to the Brønsted-Lowry definition.

$HCl \rightarrow H^+(aq) + Cl^-(aq)$	the Cl^- ion is the conjugate base of HCl
$NH_3(aq) + H^+(aq) \rightarrow NH_4^+(aq)$	the NH_4^+ ion is the conjugate acid of NH_3

8.2 Learning Check

1. Write the conjugate acid for the following bases

a) F^- *b)* N_2H_4 *c)* C_5H_5N

d) O_2^{2-} *e)* $HCrO_4^-$ *f)* HO_2^-

2. Write the conjugate base for the following acids

Definition

a) NH_3 *b)* HCO_3^- *c)* HCN

d) H_5IO_6 *e)* HNO_3 *f)* H_2O

3. Identify the conjugate pairs in the following equations

a) $HNO_3 + N_2H_4 \rightleftharpoons NO_3^- + N_2H_5^+$

b) $NH_3 + N_2H_5^+ \rightleftharpoons NH_4^+ + N_2H_4$

c) $H_2PO_4^- + CO_3^{2-} \rightleftharpoons HPO_4^{2-} + HCO_3^-$

d) $HIO_3 + HC_2O_4^- \rightleftharpoons IO_3^- + H_2C_2O_4$

Amphoteric substances

An amphoteric substance is one that may behave as either an acid or a base

An amphiprotic substance is one that may gain or lose a proton (i.e. behaves as either a Brønsted-Lowry acid or base)

Just as the Lewis definition is broader than the Brønsted-Lowry definition, the set of amphiprotic substances are contained within the set of amphoteric substances.

How do you know? – You look at the other species that it is reacting with. If it reacting with an known acid, then the amphoteric substance is behaving like a base, and vice versa.

This is a good place for IB examiners to combine questions – usually conjugates go well with amphoteric species. This is an opportunity for you to apply the definitions of acids and bases.

$$H_2CO_3(aq) \rightleftharpoons HCO_3^-(aq) + H^+(aq) \rightleftharpoons CO_3^{2-}(aq) + 2H^+(aq)$$

acid ——— | conjugate base / acid ——— | ——— conjugate base

The hydrogen carbonate ion, $HCO_3^-(aq)$, can act as a base or an acid – it's amphoteric, and because it can gain or lose a proton it's amphiprotic.

Strong & Weak

Electrolytes are any substance that dissolves in water to produce ions (and hence conduct electricity). Electrolytes may be acids, bases or salts.

A strong electrolyte dissociates completely into ions - ionizes 100%.
A weak electrolyte does not completely dissociate into ions- only partially dissociates.

	Acids	Bases
Strong	HCl - hydrochloric acid HNO_3 - nitric acid H_2SO_4 - sulphuric acid	LiOH, NaOH, KOH $Ba(OH)_2$
Weak	carboxylic acids; eg: CH_3COOH H_2CO_3 – carbonic acid– only (aq)	NH_3 – ammonia amines; eg: CH_3NH_2

Table 8.1: Some common Acids & Bases

Strong Acid: $HCl(aq) \rightarrow H^+(aq) + Cl^-(aq)$ (100% ions)
Strong Base: $NaOH(aq) \rightarrow Na^+(aq) + OH^-(aq)$

Notice the one way arrow

Weak Acid: $CH_3COOH(aq) \rightleftharpoons CH_3COO^-(aq) + H^+(aq)$ (~1% ions)
Weak Base: $NH_3(aq) + H_2O(l) \rightleftharpoons NH_4^+(aq) + Cl^-(aq)$

Notice the equilibrium arrow!

Exam Hint

Only weak acids and bases have dissociation constants, (K_a or K_b), strong acids and bases do not have dissociation constants as they fully ionize. Strong and Weak are categories of acids and bases - you cannot "weaken" a substance by adding water - that's diluting it. Don't confuse concentration with strength. Concentration is the solute / solvent ratio and can be changed by adding water (to dilute). Strength is a category of electrolyte (acid or base), and cannot be changed. Acid and bases are either strong or weak.

Category	Concentration	
	Concentrated	Dilute
Strong	12.0 M HCl(aq)	0.1 M HCl(aq)
Weak	16.0 M CH$_3$COOH	0.5 M CH$_3$COOH

Distinguishing between Strong and Weak.

In a strong acid the concentration of H$^+$(aq) is the same as the concentration stated. However, due to only partial dissociation, a weak acid has much less H$^+$(aq) than the amount stated.

For example, a 0.1 M CH$_3$COOH solution has an H$^+$(aq) concentration of only 0.00134 M - almost 100 times less!

So in "equimolar" solutions, we have much lower [H$^+$(aq)] in the weak acid compared to the strong acid.

We can do some quick chemical tests to tell the difference - based on the amount of H$^+$(aq) present. - Remember that "rate" depends on concentration.

Test	Strong Acid	Weak Acid
add reactive metal eg Mg	**fast** H$_2$ production	**slow** H$_2$ production
add a carbonate, CO$_3^{2-}$	**fast** CO$_2$ production	**slow** CO$_2$ production
conductivity	high (many ions)	low (few ions)
pH of 0.1 M solution	=1	≈ 3 - 4

Table 8.2: Chemical tests for Strong and Weak Acids

pH Scale

pH is defined as pH= –log[H$^+$(aq)]

Every step in the pH scale represents a factor of ten change in the concentration of H$^+$ ions.

pH	[H$^+$(aq)]	pOH	[OH$^-$(aq)]	[H$^+$(aq)]x[OH$^-$(aq)]	pOH + pH
1	1x10^{-1}	13	1x10^{-13}	1x10^{-14}	14
3	1x10^{-3}	11	1x10^{-11}	1x10^{-14}	14
5	1x10^{-5}	9	1x10^{-9}	1x10^{-14}	14
7	1x10^{-7}	7	1x10^{-7}	1x10^{-14}	14
9	1x10^{-9}	5	1x10^{-5}	1x10^{-14}	14
11	1x10^{-11}	3	1x10^{-3}	1x10^{-14}	14
13	1x10^{-13}	1	1x10^{-1}	1x10^{-14}	14

Table 8.3: Relationship between pH, pOH, [H$^+$(aq)] and [OH$^-$(aq)] for odd values

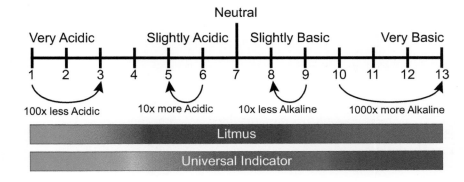

Figure 8.4: The pH Scale and Indicator colours

Acid Deposition

Carbon dioxide, a natural component of the atmosphere reacts with atmospheric moisture

$$CO_2(g) + H_2O(l) \rightarrow H_2CO_3(aq)$$

to produce carbonic acid, a weak acid with a pKa = 3.6.

This leads to rain water having a pH which is approximately 5.5. Despite being acidic, this is not considered acid rain, as rain water is always this pH. However when oxides of sulphur and nitrogen do the same thing, they drive the pH of rain water down below 5.5 - this is referred to as acid rain.

Oxides of non-metals, especially nitrogen and sulphur oxides are acidic.

Nitrogen oxides NO and NO_2 are jointly referred to as NOx. NO and NO_2 are typically produced in high temperature and pressure environments, like automobile engines.

$N_2(g) + O_2(g) \rightarrow 2NO$ $\qquad$ NO is easily oxidized by atmospheric oxygen

$2NO(g) + O_2 \rightarrow 2NO_2$ $\qquad$ NO_2 dissolves in atmospheric moisture to give acid solution

$2NO_2(g) + H_2O \rightarrow HNO_2 + HNO_3$

NO_2 is nitrous acid, a weak acid with a pK_a 3.4, while HNO_3 is nitric acid and is a strong acid.

Sulfur oxides undergo the same types of reactions

$S(s) + O_2(g) \rightarrow SO_2(g)$

$2SO_2 + O_2(g) \rightarrow 2SO_3(g)$

$SO_2(g) + H_2O \rightarrow H_2SO_3(aq)$ $\qquad$ sulphurous acid: a weak acid with pK_a 1.81

$SO_3(g) + H_2O \rightarrow H_2SO_4(aq)$ $\qquad$ sulphuric acid: a strong acid.

In order to avoid acid rain / acid deposition, we must remove the nitrogen and sulphur either before or after any combustion.

Removing sulphur oxides may be done by spraying water through the chimney and collecting the acid water for further processing to useful products. Nitrogen oxides aren't as easy.

pH Calculations

Put simply, you should always use the following three relationships to help solve problems

at any temperature $\qquad$ **at 25°C**

$K_a \times K_b = K_w$ $\qquad$ $K_a \times K_b = 1 \times 10^{-14}$

$pK_a + pK_b = K_w$ $\qquad$ $pK_a + pK_b = 14$ $\quad$ useful to find the pKa of a conjugate

$pH + pOH = K_w$ $\qquad$ $pH + pOH = 14$ $\quad$ use to switch between acid and base concentration / pHs

K_w is the equilibrium constant for water and at 25°C has the value of 1×10^{-14}.

Water is a very weak acid or base – it's amphoteric

$H_2O(l) + HCl(g) \rightarrow H_3O^+(aq) + Cl^-(aq)$ $\qquad$ water gains H^+; it acts as a base

$H_2O(l) + NH_3(g) \rightarrow OH^-(aq) + NH_4^+(aq)$ $\qquad$ water loses H^+; it acts as an acid

$K_w = [H^+(aq)] \times [OH^-(aq)] = 1 \times 10^{-14}$

Taking the negative log of both sides

Acids & Bases

$-\log[H^+(aq)] + -\log[OH^-(aq)] = -\log 1\times10^{-14}$

therefore   pH + pOH = 14

2 trick questions – the pH of a very, very diluted acid and the pH of an alkaline solution.

Exam Traps

1) What is the pH of a 1.0×10^{-9} M solution of HCl?

   wrong answer: pH = 9 ; but it's an acid, and even a diluted acid cannot be alkaline
   right answer: pH = 7 ; because the HCl is so dilute that the ionization of water is dominant

2) What is the pH of a 1.0×10^{-4} M solution of NaOH?

   wrong answer: pH = 4 you forgot to notice that we are calculating the pH of a solution containing OH⁻
   right answer: pH = 14 – pOH = 14 – 4 = 10   pH = 10 – remember it's alkaline!

pH of a strong electrolyte is dependant only on the concentration because all of the strong electrolyte dissociates into ions.

pH of a weak electrolyte depends on the concentration **and** the degree of dissociation – the K_a or K_b.

Strong Acid	Strong Base
pH = $-\log[H^+(aq)]$	pOH = $-\log[OH^-(aq)]$
	pH = 14 – pOH

Calculation of pH of weak acid solutions

In order to find the pH of a weak acid, you must first calculate the $[H^+(aq)]_{eq}$, which is usually the "x" in your "ICE" table

Determine the pH of a 0.2 M solution of ethanoic acid, pK_a = 4.74
The equation of the dissociation of ethanoic acid is

$$CH_3COOH(aq) \rightleftharpoons H^+(aq) + CH_3COO^-(aq)$$

Example

	$CH_3COOH(aq)$	$H^+(aq)$	$CH_3COO^-(aq)$
Initial	0.2 M	0	0
Change	–x	x	x
Equilibrium	0.2–x ≈ 0.2 M	x	x

The K_a is found by taking the antilog of the pKa : $K_a = 10^{-4.74} = 1.8\times10^{-5}$

Convert pK_a to K_a

$$K_a = \frac{[H^+(aq)][CH_3COO^-]}{[CH_3COOH(aq)]}$$

$$K_a = \frac{(x)(x)}{(0.2-x)}$$

$$K_a = \frac{x^2}{(0.2)}$$

Note: we make our assumption here that because x is small, we can ignore it when it is subtracted from 0.2.
So $0.2 - x \approx 0.2\ M$

$$x^2 = 0.2 \times K_a$$
$$x^2 = 0.2 \times 1.8 \times 10^{-5}$$
$$x^2 = 3.6 \times 10^{-6}$$
$$x = \sqrt{3.6 \times 10^{-6}}$$
$$x = 0.00191$$

So x= 0.00191 which represents the equilibrium value for [H⁺(aq)]

Now you calculate the pH as usual.

$$pH = -\log[H^+(aq)]$$
$$pH = -\log(0.00191)$$
$$pH = 2.72$$

8.3 Learning Check

Determine the pH of the following equimolar solutions of acids.

1. a) 0.2 M HNO₃(aq)
   b) 0.2 M HF(aq) ; $K_a = 7.2 \times 10^{-4}$

2. Find the pH of
   a) 0.05 M HCl(aq)
   b) 0.05 M HCN (aq) ; $K_a = 6.2 \times 10^{-10}$

3. Find the pH of
   a) 4×10^{-3} M HNO₃(aq)
   b) 4×10^{-3} M HOCl(aq) ; $K_a = 3.5 \times 10^{-8}$

Hydrolysis of Salts

Hydrolysis occurs when a water molecule reacts with a salt of a weak acid or base. In the case of the conjugate of a weak acid ethanoic acid, the ethanoate ion is a reasonably good base, and is able to "steal" a proton from the water to reform the weak acid, leaving behind a hydroxide ion.

$$CH_3COO^-(aq) + H_2O(l) \rightleftharpoons CH_3COOH(aq) + OH^-(aq)$$

In the case of the salt of a weak base ammonia, the conjugate contains the ammonium ion which can give away its extra proton to water forming the acidic "hydronium" ion and the original weak base.

$$NH_4^+(aq) + H_2O(l) \rightleftharpoons NH_3(aq) + H_3O^+(aq)$$

In short - the salt of a weak acid is basic, and the salt of a weak base is acidic. This will be important when we consider the titration of a weak acid or base.

The other example of hydrolysis is with highly charged (+3) ions - usually with transition metals and aluminium solutions

Recall the hexaaquairon(III) complex ion. Because of the high positive charge, one of the HO–H bonds is weakened enough to allow an H⁺(aq) ion to dissociate.

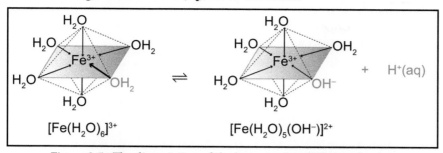

Figure 8.5: The dissociation of the hexaaquairon(III) complex ion

Buffer Solutions

It's important to distinguish between the definition, composition and preparation of buffers. Make sure you answer the question.

Definition

Definition	A solution which resists a change in pH when a **small** amount of acid or base is added.
Composition	A mixture of a weak acid and its conjugate.
	A mixture of a weak base and its conjugate.
Preparation	I – simple mixing of a weak acid or base and its conjugate.
	II – partially neutralize a weak acid (or base) with a strong base (or acid).

Compostion

Why a mixture? Well the weak acid reacts with the base to produce salts and water, but because the acid is weak, there is very little of the conjugate base around to react with any added acid. The source of the "extra" conjugate base is the salt.

The reason buffers work is equilibrium, and LeChâtelier's Principle. If you add some acid (or base), the equilibrium shifts to keep the concentration ratios constant and hence maintain the pH.

Let's look at it in terms of a chemical equilibrium

$$HA(aq) \rightleftharpoons H^+(aq) + A^-(aq)$$

If a OH^- is added it removes $H^+(aq)$ by forming water. Equilibrium shifts to the right to replace $H^+(aq)$, and there is a lot of HA around because it's a weak acid!

If H^+ is added, it shifts the equilibrium to the left by reacting with the $A^-(aq)$ to form HA. The A^- comes from the conjugate base that we added as the salt.

Preparation

I – the first method of preparation is easy – mix the weak acid/base and the salt. Usually it's in equal proportions, but it doesn't need to be.

II – the most common method of preparation of a buffer is "partial neutralization": A method is to react two parts weak acid with one part strong base. This means that there will be one part of excess weak acid left over and one part salt (conjugate) produced – therefore the mixture is created "*in situ*".

	HA(aq)	NaOH(aq)	→	Na$^+$A$^-$(aq)	H$_2$O(l)
before reaction	2	1		0	0
	XS	Limiting			
after reaction	1	0		1	1

After the reaction, you have one part weak acid (it's been only partially neutralised), and one part conjugate salt. - That's a buffer!.

Of course you can do the same thing with a weak base, and a strong acid. You need XS of the weak base and the strong acid will be the limiting reactant.

Acids & Bases

Acid – Base Titration – Strong Acid & Strong Base

When the solution is neutral (pH = 7), the number of moles must equal the number of moles of base.

$$n_{acid} = n_{base}$$

because n=CV we can substitute on both sides of the equation.

$$C_{acid} \cdot V_{acid} = C_{base} \cdot V_{base}$$

You will always know three of the four variables.

Example

What is the volume of 0.625 M NaOH(aq) is required to neutralize 20.00 cm³ of 0.125 M HCl?

Solution

$C_{acid} \cdot V_{acid} = C_{base} \cdot V_{base}$
$V_{base} = C_{acid} \cdot V_{acid} \div C_{base}$
$V_{base} = 0.125\ M \cdot 20.00\ cm^3 \div 0.625\ M$
$V_{base} = 40.0\ cm^3$

Note: you do not have to convert to dm³, because the concentration units will cancel out.

Example

What is the volume of 0.625 M NaOH(aq) is required to neutralize 20.00 cm³ of 0.125 M H₂SO₄?

Solution - slightly more complicated because we have a diprotic acid...

$$H_2SO_4(aq) + 2NaOH \rightarrow Na_2SO_4(aq) + 2H_2O(aq)$$

Because we need twice as much base as acid

$n_{base} = 2 \times n_{acid}$
$C_{base} \cdot V_{base} = 2(C_{acid} \cdot V_{acid})$
$V_{base} = 2(C_{acid} \cdot V_{acid} \div C_{base})$
$V_{base} = 2 \times (0.125\ M \cdot 20.00\ cm^3 \div 0.625\ M)$
$V_{base} = 80.00\ cm^3$

this makes sense, because we need double the volume compared to previous question because there is twice as many H⁺(aq) ions in the acid.

Example

What is the pH of the resulting solution when 20.00 cm³ of 0.10 M NaOH is added to 100.00 cm³ of 0.20 M HCl(aq)?

- moles of NaOH = 0.02000 dm³ × 0.1 mol dm⁻³ = 0.0020 mol NaOH
- moles of HCl = 100.00 dm³ × 0.20 mol dm⁻³ = 0.02 mol HCl

- Stoichiometric ratio is 1:1, NaOH is Limiting, 0.0020 mol reacted.
  0.020 mol - 0.002 mol = 0.018 mol of HCl unreacted.
- Total volume = 20.00 cm³ + 100.00 cm³ = 120.00 cm³ = 0.120 dm³. Don't forget to add the volumes!!
- Concentration of HCl(aq) after = 0.018 mol ÷ 0.120 dm³
  = 0.15 mol dm⁻³ = 0.15 M

- pH = -log[H⁺(aq)]
- pH = log(0.15 M)
- pH = 0.82

Acid – Base Titration – Weak & Strong

A lot of students are still very excited about equilibrium, so they forget that in a neutralization reaction the formation of water is a very strong driving force, and therefore the neutralization reaction is **not an equilibrium**.

Equilibrium exists "after" the reaction has occurred.

In a titration question you must obey the stoichiometry first, then look for the excess reactant and the product formed. One of the species will be the limiting reactant. You can use an "ICE" table or the Henderson-Hasselbalch equation to find the resulting pH.

Example

What is the pH of the resulting solution when 20.00 cm³ of 0.10 M NaOH is added to 100.00 cm³ of 0.20 M $CH_3COOH(aq)$? The pK_a of CH_3COOH is 4.74.

Solution

We first need to determine the concentrations of all species after the reaction has occurred.

$$CH_3COOH(aq) + NaOH(aq) \rightarrow Na^+(aq) + CH_3COO^-(aq) + H_2O(l)$$

NaOH is limiting reactant

n_{NaOH} = 0.020 dm³ × 0.1 mol dm⁻³ = 0.0020 mol NaOH

$n_{acid\ unreacted}$ = (0.100 dm³ × 0.20 mol dm⁻³) − 0.0020 mol = 0.018 mol CH_3COOH

$n_{salt\ formed}$ = n_{NaOH} = 0.0020 mol

Don't forget to add the volumes when determining the concentrations.

[acid] = 0.018 mol ÷ 0.120 dm⁻³ = 0.15 M

[salt] = 0.002 mol ÷ 0.120 dm⁻³ = 0.0167 M

Secondly, we apply the equilibrium concept and the Henderson-Hasselbalch equation

pH = pKa + log ([salt] ÷ [acid])

pH = 4.74 + log (0.0167 ÷ 0.15)

pH = 4.74 − 0.953

pH = 3.79

This result makes sense, because we have more acid than salt, so it drives the pH downwards from the pK_a.

8.4 Learning Check

What is the pH of the resulting solution when 20.00 cm³ of 0.20 M HCl is added to 100.00 cm³ of 0.10 M $NH_3(aq)$? The pK_b of NH_3 is 4.75.

Indicators

An indicator is a weak acid or base that has different colours in different pHs

An indicator is represented generically by HIn - it is a special case of HA, the weak acid.

$$HIn \rightleftharpoons H^+(aq) + In^-(aq)$$

colour 1   colour 2

Adding $H^+(aq)$ will shift the equilibrium to the left and colour 1. Whereas addition of $OH^-(aq)$ will remove $H^+(aq)$, and shift the equilibrium to the right and colour 2.

The pK_a of the indicator gives us the pH where the colour change is centered. See your data booklet - Table 22.

pH Curves

When sketching pH curves there are two key landmarks to consider.

1. The initial pH - this is the pH of the original solution.
   A 0.1M strong acid will have an initial pH of 1, 0.1 M strong base starts at pH 13.
   A 0.1 M solution of a weak acid will have an initial pH above 1.
   A 0.1 M solution of a weak base will have an initial pH below 13.

2. The pH of the **equivalence** point. (It's not necessarily neutral!!!)
   A strong acid/strong base will have an equivalence point of pH = 7
   A weak acid, strong base will have an equivalence point of pH>7.
   A weak base, strong acid will have an equivalence point of pH<7.

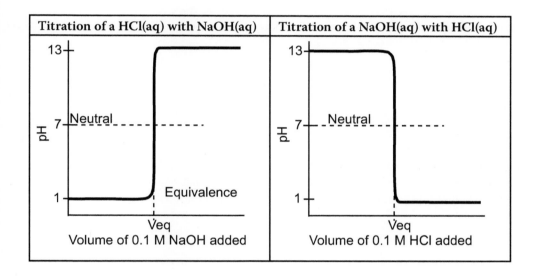

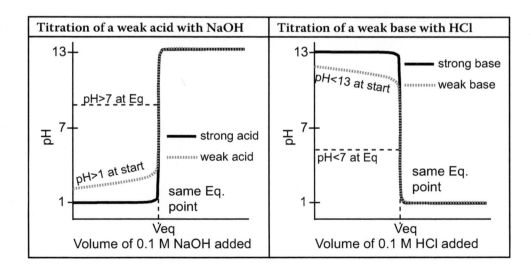

Relative Strengths of Acids & Bases

	Acid	K_a	pK_a	Conjugate base	K_b	pK_b	
strong acid	HCl	v. high		Cl$^-$			**weak base**
	CCl$_3$COOH	2.2×10^{-1}	0.65	CCl$_3$COO$^-$		13.35	
	CH$_2$ClCOOH	1.38×10^{-3}	2.86	CH$_2$ClCOO$^-$		11.14	
	HCOOH	1.77×10^{-4}	3.75	HCOO$^-$		10.25	
	CH$_3$COOH	1.74×10^{-5}	4.76	CH$_3$COO$^-$		9.24	

	Conjugate acid			Base			
weak acid	NH$_4^+$	5.62×10^{-10}	9.25	NH$_3$		4.75	**strong base**
	CH$_3$NH$_3^+$	2.29×10^{-11}	10.64	CH$_3$NH$_2$		3.36	
	H$_2$O	—	—	NaOH	—	strong	

Table 8.6: Relative Strengths of Acids & Bases

Looking at Table 8.7, we can see that as the acid strength increases, the strength of the conjugate base decreases. Therefore strong acids have very weak conjugate bases, and weak acids have reasonably strong conjugate bases.

Look at the equilibrium reaction for ammonia and water.

$$NH_3(aq) + H_2O(l) \rightleftharpoons NH_4^+(aq) + OH^-(aq)$$

The K_b value is really a measure of which base, NH$_3$, or OH$^-$ is stronger. We know that the hydroxide ion is a very strong base, and we confirm that by seeing that a small K_b means that the equilibrium lies to the reactant side.

Acids & Bases

Summary Questions.

1. Write balanced equations for
    a) nitric acid and copper(II) oxide
    b) aluminium and hydrochloric acid
    c) iron(III) carbonate and sulphuric acid

2. Write the conjugate acid for the following bases.
    a) SO_4^{2-} b) CO_3^{2-} c) $CH_3CO_2^-$
    d) NH_2^- e) NH_3 f) HPO_4^{2-}

3. Write the conjugate base for the following acids.
    a) HI   b) H_2 c) NH_4^+
    d) HNO_2 e) $H_2PO_4^-$ f) H_3PO_4

4. Identify the conjugate pairs in the following equations.
    a) $HSO_4^- + SO_3^{2-} \rightleftharpoons HSO_3^- + SO_4^{2-}$
    b) $S^{2-} + H_2O \rightleftharpoons HS^- + OH^-$
    c) $CN^- + H_3O^+ \rightleftharpoons HCN + H_2O$
    d) $H_2Se + H_2O \rightleftharpoons HSe^- + H_3O^+$

5. Determine the pH of the following solutions.
    a) 0.1 M HCl
    b) 1×10^{-3} HNO_3
    c) 1×10^{-8} HCl
    d) 0.1 M NaOH
    e) 1×10^{-9} NaOH

C.Lumsden   IB HL Chemistry

Chapter 9

Oxidation & Reduction

In this chapter...

- 138 Introduction
- 138 Oxidation numbers.
- 139 Redox Reactions
- 140 Reactivity
- 140 Half Reactions
- 141 Dissolved Oxygen and BOD
- 141 Definitions
- 142 Voltaic Cells
- 142 Salt Bridge
- 143 Standard Electrode Potentials
- 144 Electrolysis of molten salts
- 144 Electrolysis of solutions
- 145 Factors affecting amount of electrolysis product
- 146 Balancing Redox reactions in Acidic Solution
- 147 Chapter 9 Summary questions

Introduction

Oxidation & Reduction - *"Redox"* to chemists - reactions are those that involve a transfer of electrons. Don't all reactions involve a transfer of electrons? – NO! Acid/Base reactions and precipitation reactions do not involve a transfer of electrons.

Definition

	Oxidation	Reduction
Oxygen	Gain of oxygen	loss of oxygen
Electrons	**Loss of electrons**	**Gain of electrons**
Hydrogen	Loss of hydrogen	gain of hydrogen
Electrode	**Anode**	**Cathode**
oxidation number	becomes more positive	becomes less positive

In IB Chemistry, we are mostly concerned with the electronic definition of oxidation and reduction. The oxygen and hydrogen are included as additional information. - The hydrogen definition should be familiar to biologists.

Exam Hint

My favourite way to remember this is **"AN OIL RIG CAT"** - **AN**ode is the electrode where **O**xidation **I**s **L**oss of electrons occurs, and **R**eduction **I**s **G**ain of electrons at the **CAT**hode.

Oxidation numbers.

Oxidation numbers are chemists' way to keep an account of electrons.

> The oxidation number of any free element is zero, regardless of how complex its molecules might be.
>
> The oxidation number of any simple, monatomic ion is equal to the charge on the ion.
>
> The sum of all the oxidation numbers of the atoms in a molecule is equal to the charge on the particle (neutral for molecules, ionic charge for complex ions.)
>
> In its compounds, group I and II metals have an oxidation number of +1 and +2 respectively.
>
> In its compounds, fluorine has an oxidation number of –1.
>
> In its compounds, hydrogen usually has an oxidation number of +1 (except $LiAlH_4$ H= –1)
>
> In its compounds, oxygen has an oxidation number of –2. (Except the peroxide ion: O_2^{2-})
>
> * When two rules conflict, take the rule with the lower number.

Students should be organized in their working for determining oxidation numbers.

Write the oxidation number for "each" atom on the top.

Write the total contribution for that element below.

$$\text{each } \overset{+6 \;\; -2}{Cr_2O_7^{2-}}$$
$$\text{total } +12 \; -14 = -2$$

Oxidation & Reduction

It's the bottom numbers that have to add to zero (or the overall ionic charge).
Determine the oxidation number (state) of the element in **bold** type.

9.1 Learning Check

MoS_2 Ni_2**O**$_3$ **P**$_4O_6$ **As**$_2O_3$

Cr$(NO_3)_3$ Cr_2(**S**$O_4)_3$ **Cr**SO_4 Cr(S**O**$_4)_3$

ClO^- **Cl**O_2^- **Cl**O_3^- **Cl**O_4^-

Redox Reactions

A redox reaction is one that has changes in oxidation number. There are lots of reactions which are not redox reactions – precipitations, acid/base reactions etc.

Reactions that are always going to be redox reactions:

- Combustion reactions – elemental oxygen becoming a compound.
- Synthesis reactions – any element reacting with another to produce a compound.
- Ions changing charge – Fe^{2+} becoming Fe^{3+}.
- Ions of the oxy-acids changing the number of oxygen atoms – SO_3^{2-} becoming SO_4^{2-}.

A fast way to pick off 99% of redox reactions is to look for an element (O_2, Cl_2, Fe, Na etc) as a reactant or product. This works because the oxidation state of any free element is zero, and in a compound, it is unlikley zero.

Exam Hint

Determine if the following reactions are oxidation-reduction reactions by determining any changes in oxidation number.

9.2 Learning Check

1. $H_2 + Cl_2 \rightarrow 2HCl$
2. $2KCl + MnO_2 + 2H_2SO_4 \rightarrow K_2SO_4 + MnSO_4 + Cl_2 + 2H_2O$
3. $CH_4(g) + 2O_2(g) \rightarrow CO_2(g) + 2H_2O(g)$
4. $Cr_2O_7^{2-}(aq) + 2OH^-(aq) \rightarrow 2CrO_4^{2-}(aq) + H_2O(aq)$
5. $Al(OH)_4^-(aq) \rightarrow AlO_2^-(aq) + 2H_2O(l)$

Oxidizing agent – a species that removes electrons from another.

Reducing agent – a species that donates electrons to another.

Common Oxidizing Agents	Common Reducing Agents
MnO_4^- (acidic, basic or neutral)	$NaHSO_3$
$Cr_2O_7^{2-}$ (acidified)	$LiAlH_4$
H_2O_2 (acidic, basic or neutral)	$S_2O_3^{2-}$

Table 9.1: Some common Oxidizing & Reducing Agents

Identify the oxidizing and reducing agents from the previous learning check.

9.3 Learning Check

C.Lumsden   IB HL Chemistry

Reactivity

Reactivity always compares elements to each other, not their ions.

Reactivity of the Halogens - See Chapter 3 for more details.

Metals have an order of reactivity that you are not expected to memorise, but rather you may be asked to deduce an order of reactivity given some reactions.

Example

Determine the order of reactivity given the following reactions.

$$Mg(s) + Zn^{2+}(aq) \rightarrow Mg^{2+}(aq) + Zn(s)$$
$$Fe(s) + Cu^{2+}(aq) \rightarrow Cu(s) + Fe^{2+}(aq)$$
$$Zn(s) + Fe^{2+}(aq) \rightarrow Zn^{2+}(aq) + Fe(s)$$

Mg is more reactive than Zn.
Fe is more reactive than Cu.
Zn is more reactive than Fe.

Therefore the metals in order of their reactivity is Mg>Zn>Fe>Cu

Half Reactions

Because electrons are transferred, for any oxidation, something must be reduced, and vice versa. We can write each of these processes separately, but including the electrons

reduction half reaction: $\quad Cl_2 + 2e^- \rightleftharpoons 2Cl^-$

oxidation half reaction: $\quad\quad\quad Na \rightleftharpoons Na^+ + e^-$

Just like in Hess' law, we need to add these two half reactions together to make a whole reaction. The trick I that the electrons must cancel out, so we need to double the sodium half reaction.

Reduction	$Cl_2(g)$ +2e⁻	$\rightleftharpoons$ 2Cl⁻
Oxidation	2Na(g)	$\rightleftharpoons$ 2Na⁺(g) +2e⁻
Net reaction	2Na(g) + $Cl_2(g)$	$\rightarrow$ 2NaCl(s)

Get the point

If you are writing the half reaction, use the equilbrium sign. However in the full chemical reaction, there is a strong driving force pushing the reaction in the spontaneous direction. Therefore don't use the double equilibrium arrow in the reaction.

Dissolved Oxygen and BOD

The health of a river, lake or ocean, may be measured by how much oxygen and oxygen consuming organisms are present.

In order to determine the amount of oxygen dissolved in the water, we take advantage of the fact that the transition metal manganese has multiple oxidation states.

1) $2Mn^{2+}(aq) + 4OH^-(aq) + O_2(aq) \rightarrow 2MnO(OH)_2(s)$

2) $MnO(OH)_2(s) + 4H^+(aq) + 2I^-(aq) \rightarrow I_2(aq) + Mn^{2+}(aq) + 2H_2O$

3) $I^-(aq) + I_2(aq) \rightarrow I_3^-(aq)$

4) $I_3^-(aq) + 2S_2O_3^{2-}(aq) \rightarrow 2I^-(aq) + S_4O_6^{2-}(aq)$

In reaction 1, the manganese ion is being oxidized from +2 to +4 by the oxygen under alkaline conditions. This "fixes" the oxygen, so that it is now represented by the manganese complex. Typically this is done in a small air tight flask which is filled to the very top to exclude any interfering oxygen in the air. In the second reaction the $MnO(OH)_2$ is oxidizing the iodide ions to iodine the react with the excess iodide to form the triiodide ion, $I_3^-(aq)$. The $I_3^-(aq)$ can be titrated in the fourth reaction by thiosulphate. The result of the titration allows us to determine the original concentration of O_2 in the water.

Typically a large sample is collected and part of it is analysed immediately. If a second portion of the sample is analysed 5 days later we can determine the amount of oxgyen consumed by aerobic bacteria. The difference between the two readings is called Biologlical Oxygen Demand for 5 days or BOD5. If the BOD5 is low, then there isn't much microbial oxygen demand. This means the oxygen is there for larger organisms like fish. However if the BOD5 is high, then the water is polluted with oxygen demanding microbes and the fish etc. will not be able to survive.

Typically Dissolved Oxygen (DO) is measured in mg/dm³ of water - this is the same as parts per million (ppm).

Gas solubility decreases with temperature - so we find larger fish in colder, deeper lakes where the DO can be high enough. Fish are cold blooded organisms, so their metabolic rate increases when the temperature is higher as well. Therefore they need even higher DO levels if the water is warm.

Eutrophication is a contributor to high BOD as untreated sewage and fertilizers contribute to plant respiration despite any photosynthetic processes occurring.

Definitions

There are several ways to view oxidation and reduction:

Oxidation	Reduction
gain of oxygen	loss of oxygen
loss of electrons	gain of electrons
loss of hydrogen	gain of hydrogen
increase in oxidation number	decrease in oxidation number

Gain and loss of oxygen seems obvious enough. When gaining oxygen, the other atom tends to "give" the more electronegative oxygen its electrons.

Voltaic Cells

A voltaic cell allows a chemical reaction to produce electricity based on difference in reactivity. The greater the difference in reactivity the greater the potential difference (voltage).

If one were to place a reactive metal into a solution of a less reactive metal, the electron transfer would occur at the interface of the two materials – i.e. If you dipped a zinc rod into a solution of copper(II) ions, the electrons in the zinc metal would transfer directly to the copper(II) ions and therefore cannot be harnessed to do any useful work.

In a Voltaic (also known as a Galvanic Cell), you must keep the two half reactions separate so that the electrons are forced to travel through an external circuit where they can do useful work.

The **anode** is defined as the electrode where oxidation occurs.

The **cathode** is defined as the electrode where reduction occurs.

Cell Drawing Check List
- electrodes labelled with metal type
- electrodes immersed in solution of metal ions
- external circuit
- salt bridge (in solutions)
- half reactions
- anode / cathode labelled
- electrode polarity labelled
- electron flow in external circuit shown

- you don't have to draw in 3-D though!

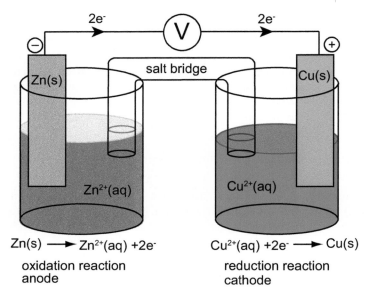

Figure 9.2: The Zinc/Copper Cell

Observations: The zinc electrode will gradually dissolve / lose mass as the solid is oxidized. The copper solution will lose it's blue colour (fades to colourless), and the copper electrode gains mass as copper ions are deposited.

Salt Bridge

You may explain the purpose of the salt bridge in any of the following ways...
- The salt bridge completes the circuit.
- The salt bridge allows for the conduction of **ions**.
- The salt bridge allows for the balancing of ionic charge.

Common Mistake: Do not say that *electrons* flow through the salt bridge. – NO, NO, NO.

ELECTRONS flow through the wire in the external circuit.

IONS flow through the salt bridge.

Oxidation & Reduction

Standard Electrode Potentials

Have a look at the Standard Electrode potentials in the data book. There are some important things to note down in your review...

- All reactions are written as **reductions**.
- The standard hydrogen electrode (SHE) is defined as 0.00 V (something had to be).
- There are two important reactions of water – **highlight them**. They occur at –0.83 V and +1.23 V. You will refer to these later.
- There are three different reactions of copper – make sure that you pick the right one!!!
- Obey significant figures when working out cell potentials – keep 2 decimal places.

Exam Hints

The values for E° are not affected by the co-efficients in the balanced equation - DO NOT MULTIPLY BY MOLE RATIOS!

Common Mistake

Because all reactions are written as reductions, one of the reactions must be reversed so that there is an oxidation half reaction. Which one?

You need to reverse the top reaction so that the sum of the reactions (similar to the Hess' Law calculations) becomes a positive number for the overall reaction.

Positive cell values are spontaneous. This is explained by the equation:

$$\Delta G° = - nFE°$$

Where "n" is the number of moles of electrons in the balanced equation, F is Faraday's constant (96500 C mol⁻¹), and E° is the overall cell potential. This means that a positive $E°_{cell}$, will give a negative ΔG which is required for a spontaneous reaction, as we learned in "Energetics"

Recall that a Volt is really a Joule per Coulomb (1 V = 1 J•C⁻¹)

Determine the cell potential and $\Delta G°$ for the cell...

$$Cu^{2+}(aq) + Mg(s) \rightleftharpoons Cu(s) + Mg^{2+}(aq)$$

Example

The data booklet gives values for reduction half-reactions, so we need to reverse the equation and the sign of the electrode potential for the magnesium oxidation half-reaction

Reduction half-reaction $\quad Cu^{2+}(aq) + 2e^- \rightleftharpoons Cu(s) \quad\quad +0.34$ V
Oxidation half-reaction $\quad Mg(s) \rightleftharpoons Mg^{2+}(aq) + 2e^- \quad\quad +2.37$ V

Overall: $\quad Cu^{2+}(aq) + Mg(s) \rightleftharpoons Cu(s) + Mg^{2+}(aq) \quad +2.71$ V

$\Delta G = -2 \text{ mol} \times 96500 \text{ C mol}^{-1} \times 2.71 \text{ JC}^{-1}$
$\quad\quad = -523030$ J
$\quad\quad = -523$ kJ

Use your data book to determine the cell potentials for the spontaneous reactions

a) $Zn(s) + Pb^{2+}(aq) \rightarrow Zn^{2+}(aq) + Pb(s)$
b) $Fe(s) + Ag^+(aq) \rightarrow Fe^{2+}(aq) + Ag(s)$
c) $Cu^{2+}(aq) + Zn(s) \rightarrow Cu(s) + Zn^{2+}(aq)$

9.4 Learning Check

Use your data booklet to determine if the following reactions are spontaneous or not.

a) $Pb(s) + Fe^{2+}(aq)$
b) $Br_2(aq) + 2Cl^-(aq)$
c) $Cu(s) + H^+(aq)$
d) $Mg(s) + H_2O$
e) which of $Cr_2O_7^{2-}$ or $MnO_4^-(aq)$ can oxidize chloride ions?

9.5 Learning Check

Electrolysis of molten salts

An electrolytic cell uses a potential difference (electricity) to force (an otherwise non-spontaneous) reaction to occur. This is opposite to the process occuring in voltaic cells)

When a molten (melted) salt is electrolysed, the positive ion (cation) migrates towards the negative electrode where it is discharged by reduction by gaining electrons from the electrode (cathode).

The negative ion (anion) migrates towards the positive electrode where it is discharged by oxidation by losing electrons to the electrode (anode)

Because there are no other elements around, and salts always have a metal and non-metal, it's easy to figure out the product.

Be Careful

Candidates often forget to write the correct formula of the elements - remember that halogens, hydrogen and oxygen are diatomic.

Wrong answer
$NaCl(l) \rightarrow Na + Cl$

Correct Answer
$2NaCl(l) \rightarrow 2Na(l) + Cl_2(g)$
Use state symbols too!

Electrolysis of solutions

When a solution is electrolysed, we must now consider the presence of water. Remember how you highlighted the two water reactions in the previous section (at -0.83 V and $+1.23$ V)? Now it's time to use them.

It's easy. These are the two "boundaries" that represent the electrolysis of water. Simply, water is more easily reduced at -0.83 V than anything above it. If your solution contains cations with a more negative value (above in the table), then you get $H_2(g)$ instead from the reduction of water.

Similarly, if your solution contains anions with a more positive value than $+1.23$ V (below in the table), then you obtain $O_2(g)$ from the oxidation of water.

IB Chemistry considers that most complex ions (SO_4^{2-}, NO_3^- etc.) are stable towards oxidation, and therefore water is oxidized instead to give $O_2(g)$.

Special Exception

Chlorides. A dilute solution of chloride ions will generate oxygen at the anode (as expected from the rules above). Usually we deal with standard (1.0 M) solutions. However, when dealing with concentrated solutions we can expect some variation. A concentrated solution of NaCl(aq) (called "brine") will yield chlorine gas, $Cl_2(g)$ instead.

9.6 Learning Check

Follow the example and determine the products at each electrode for the following electrolytes.

Electrolyte	Cathode Product	Anode Product
NaCl(l)	Na(l)	$Cl_2(g)$
NaCl(aq) dilute	$H_2(g)$	$O_2(g)$
NaCl(aq) conc.	$H_2(g)$	$Cl_2(g)$
$CuSO_4$(aq)		$O_2(g)$
Na_2SO_4(aq)		
$MgBr_2$(aq)		
$PbBr_2$(l)		
$PbBr_2$(aq)		

Factors affecting amount of electrolysis product

Electrolysis is the passing of an electric current through an electrolyte to cause species to be oxidized or reduced.

Electric current is the amount of charge (number of electrons) flowing per time. So having a higher current or running the electrolysis for longer means that we will obtain more product

Consider the reduction of sodium ions:

$$Na^+ + e^- \rightarrow Na$$

We need one mole of electrons to react with one mole of sodium ions to produce one mole of product.

However if we consider the reduction of magnesium...

$$Mg^{2+} + 2e^- \rightarrow Mg$$
$$Al^{3+} + 3e^- \rightarrow Al$$

We need twice as many moles of electrons to reduce one mole of magnesium. If we used the same number of electrons as in the sodium reaction, we would end up with half the amount of magnesium.

We will produce ⅓ the amount of aluminium as sodium.

Factor	Result
increase current	more product
increase time	more product
charge on ion	the greater the charge, the more electrons per reduction, therefore the less product formed

Balancing Redox reactions in Acidic Solution

Firstly you need to write the separate equations for the oxidation and reduction half reactions. For each reaction...

1. Balance all the elements except hydrogen and oxygen.
2. Balance oxygen using H_2O.
3. Balance hydrogen by using $H^+(aq)$.
4. Balance the charge by adding electrons.
5. If necessary, multiply the half reactions to equalize the number of electrons (cross multiply).
6. Add the half reactions and cancel identical species (e⁻s and H_2O).
7. Check elements and charges are balanced.

Example Balance the following reaction

$$Cr_2O_7^{2-}(aq) + Cl^-(aq) \rightarrow Cr^{3+}(aq) + Cl_2(g)$$

Reduction reaction	$Cr_2O_7^{2-}(aq) + \rightarrow Cr^{3+}(aq)$
Balance Cr	$Cr_2O_7^{2-}(aq) + \rightarrow 2Cr^{3+}(aq)$
Balance oxygen	$Cr_2O_7^{2-}(aq) + \rightarrow 2Cr^{3+}(aq) + 7H_2O(l)$
Balance hydrogen	$14H^+(aq) + Cr_2O_7^{2-}(aq) + \rightarrow 2Cr^{3+}(aq) + 7H_2O(l)$
Balance charge	$6e^- + 14H^+(aq) + Cr_2O_7^{2-}(aq) + \rightarrow 2Cr^{3+}(aq) + 7H_2O(l)$

Oxidation reaction	$Cl^-(aq) \rightarrow Cl_2(g)$
Balance Cl	$2Cl^-(aq) \rightarrow Cl_2(g)$
Balance O	not applicable
Balance H	not applicable
Balance charge	$2Cl^-(aq) \rightarrow Cl_2(g) + 2e^-$

Now we need to multiply the oxidation reaction by 3 to make the electrons in both the reactions the same. Add the equations.

$$6e^- + 14H^+(aq) + Cr_2O_7^{2-}(aq) + \rightarrow 2Cr^{3+}(aq) + 7H_2O(l)$$
$$6Cl^-(aq) \rightarrow 3Cl_2(g) + 6e^-$$

$$14H^+(aq) + Cr_2O_7^{2-}(aq) + 6Cl^-(aq) \rightarrow 2Cr^{3+}(aq) + 7H_2O(l) + 3Cl_2(g)$$

You can see from this equation, why we always say "acidified" potassium dichromate - the oxidizing agent requires it!

9.7 Learning Check

Balance the following reactions in acidic solution.

a) $Zn(s) + HCl(aq) \rightarrow Zn^{2+}(aq) + H_2(g)$
b) $I^-(aq) + ClO^-(aq) \rightarrow I_3^-(aq) + Cl^-(aq)$
c) $As_2O_3(s) + NO_3^-(aq) \rightarrow H_3AsO_4(aq) + NO(g)$
d) $Br^-(aq) + MnO_4^-(aq) \rightarrow Br_2(l) + Mn^{2+}(aq)$
e) $CH_3OH(aq) + Cr_2O_7^{2-}(aq) \rightarrow CH_2O(aq) + Cr^{3+}(aq)$

Chapter 9 Summary questions

1. Determine the oxidation numbers of the elements in bold type.

   CO   **C**O$_2$ **Hg**$_2$Cl$_2$ **Hg**O

   K**Mn**O$_4$ Mg$_2$**P**$_2$O$_7$ **Xe**OF$_4$ **As**$_4$

   Na$_2$**C**$_2$O$_4$ Na$_2$**S**$_2$O$_3$ H**As**O$_2$ **S**$_8$

2. Balance the following acidic oxidation – reduction reactions.
   a) $H_2O_2 + Fe^{2+} \rightarrow Fe^{3+} + H_2O$
   b) $S_2O_3^{2-} + I_2 \rightarrow I^- + S_4O_6^{2-}$

3. Determine whether the following reactions are spontaneous. If they are not, re-write the spontaneous equation – determine the cell potential for the spontaneous reaction.
   a) $Cu^+(aq) + Fe^{3+}(aq) \rightarrow Cu^{2+}(aq) + Fe^{2+}(aq)$
   b) $2Br^-(aq) + Sn^{2+}(aq) \rightarrow Sn(s) + Br_2(aq)$
   c) $Ni^{2+}(aq) + Ag(s) \rightarrow Ni(s) + Ag^+(aq)$

Chapter 10

Organic Chemistry

In this chapter...

150	Introduction
150	Hydrocarbons
151	Naming Hydrocarbons.
153	Isomers
154	Structural Isomers
154	Branching in Organic Chemistry
155	Alkanes
156	Organic Functional Groups
158	Functional Groups vs. Classes of compounds
158	Functional Group Isomerism
159	Alkenes — Electrophilic Addition Reactions
162	Electrophilic Addition Mechanism
162	Benzene and Derivatives
163	Important isomers
164	Halogenoalkanes
164	Nucleophilic Substitution reactions
165	S_N1 – Nucleophilic Substitution – First order kinetics
166	S_N2 – Nucleophilic Substitution – Second order kinetics
167	Nucleophilic Substitution Reactions II
168	Alcohols
168	Combustion of alcohols
168	Oxidation of alcohols
169	Alcohol Oxidation Products
170	Reduction of aldehydes, ketones and acids
170	Functional Groups containing Nitrogen
172	Elimination Reactions
173	Ester Condensation Reactions
174	Naming Esters
174	Amide Condensation Reactions
175	Condensation Polymers
177	Electrophilic Substitution Reactions
177	Reaction Pathways
178	Stereoisomerism
179	Optical Isomerism
180	Summary Questions

Introduction

Organic chemistry is the study of carbon compounds. Carbon has the unique ability to form chains (concatenate). Because these chains can have branches and other atoms present, there are more organic chemicals than all other combined.

Carbon has a valence of four - that is, it always makes four bonds. Despite how simple this is, there are many students who make simple avoidable mistakes.

Hydrocarbons

Hydrocarbons are compounds that contain only carbon and hydrogen (not water - that's a carbohydrate!). If a hydrocarbon is bonded to the maximum number of hydrogens, then we say it is saturated. Saturated hydrocarbons have only single carbon-carbon bonds.

Hydrocarbon chains may be straight or branched, and may or may not contain double or triple bonds.

	Alkanes	Alkenes	Alkynes
General Formula	C_nH_{2n+2}	C_nH_{2n}	C_nH_{2n-2}
Bonding	single bonds	at least one double bond	at least one triple bond
Hybridization*	sp^3	sp^2	sp
Bond Geometry	tetrahedral	planar triangle	linear
C-4 example	butane C_4H_{10}	but-2-ene C_4H_8	but-2-yne C_4H_6
Saturation	saturated	unsaturated	unsaturated

Table 10.1: Hydrocarbon properties

* The hybridization of sp^2 and sp and their corresponding geometries only applies to the carbons involved in the double or triple bond. There may be sp^3 hybridized atoms elsewhere in the molecule.

Homologous Series

A homologous series is a group of compounds whose **successive** members differ by CH_2.

(Note: <u>not all</u> members differ by CH_2, just the ones next to each other.)

n	Name	Molecular formula	Condensed structural formula
1	methane	CH_4	CH_4
2	ethane	C_2H_6	CH_3-CH_3
3	propane	C_3H_8	CH_3-CH_2-CH_3
4	butane	C_4H_{10}	CH_3-CH_2-CH_2-CH_3
5	pentane	C_5H_{12}	CH_3-CH_2-CH_2-CH_2-CH_3
6	hexane	C_6H_{14}	CH_3-CH_2-CH_2-CH_2-CH_2-CH_3

Table 10.2: Alkane homologous series

Organic Chemistry

A homologous series has the same "general formula" (don't confuse this with empirical formula). In the case of the alkane series above, it is C_nH_{2n+2}. Any compound that has 2n+2 hydrogens to carbons is recognized as being saturated.

What's important to remember is that homologous series have similar chemical properties and display a gradual change in physical properties

Name	Formula	Boiling point (°C)	Density (g/cm³)	ΔH_{vap} (kJ/mol)
methane	CH_4	-161.6	0.000717	8.16
ethane	C_2H_6	-88.6	0.001212	14.64
propane	C_3H_8	-42.1	0.001830	18.71
butane	C_4H_{10}	-0.5	0.002480	22.36
pentane	C_5H_{12}	36.1	0.626000	25.70
hexane	C_6H_{14}	69.0	0.655000	27.10

Table 10.3: Physical Property trends in Alkanes

Remember that physical properties are governed by the intermolecular forces (hydrogen bonding, dipole-dipole attraction and London dispersion forces). As the hydrocarbon chain becomes longer, the London forces increase. There still also may be hydrogen bonding (acids & alcohols) or dipole interactions (esters, ketones, aldehydes).

As the chain increases...

Exam Hint

- melting point increases for all compounds - due to increased London forces.
- solubility of in water decreases for alcohols and acids, because the proportion of hydrogen bonding in the molecule decreases.

Naming Hydrocarbons.

The names of hydrocarbons can be intimidating, but there is a very simple system.

1. Pick the **longest continuous chain** of carbon atoms and obtain its root name. This is not necessarily horizontally or left to right!!!!

2. Number the chain (either from left to right or vice-versa) in order to have the **lowest possible numbers** for the different attached hydrocarbon groups.

3. Name the hydrocarbon groups attached to the longest chain by changing the suffix -ane of the root name to - yl and indicate the point of attachment by the number of the carbon atom to which the group is attached. Common group names are:

    Methyl (-CH_3); ethyl (-CH_2CH_3)

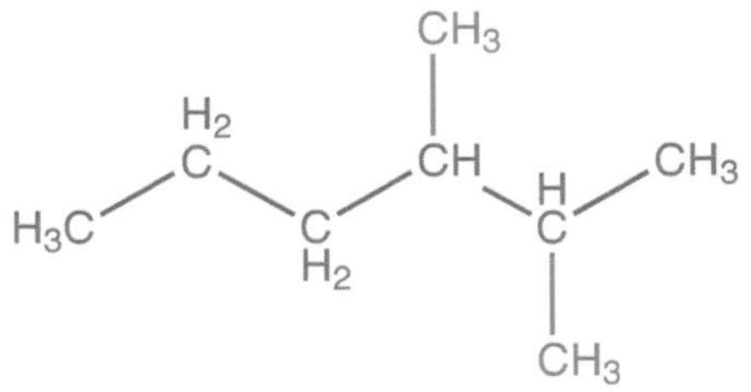

Figure 10.4: 2,3-dimethylhexane

There are many possible choices for branched groups - they could be alkyl chains (methyl, ethyl etc), or halogens (chloro, bromo, iodo)

> When two or more branches are present, you...
> Name them alphabetically
> Indicate the number of the same group by di, tri, etc, (not a factor in the alphabetical naming)
> Indicate the positions of the groups with numbers.
> Use comma's to separate numbers, and hyphens(-) to separate numbers and letters.

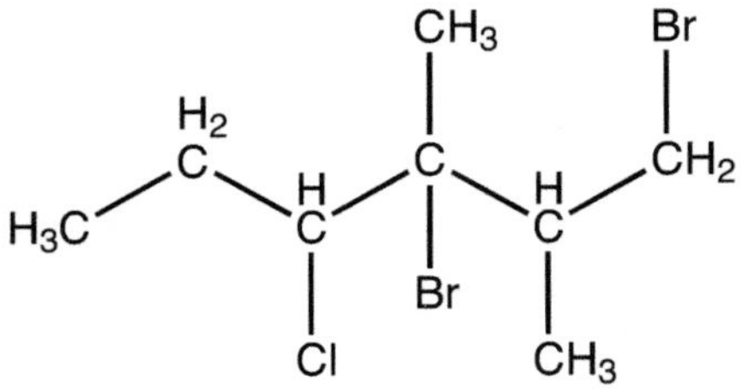

Figure 10.5: 1,3-dibromo-4-chloro-2,3-dimethylhexane

10.1 Learning Check

Select the longest carbon chain and rewrite so that the longest chain is horizontal. Write the name of the compound.

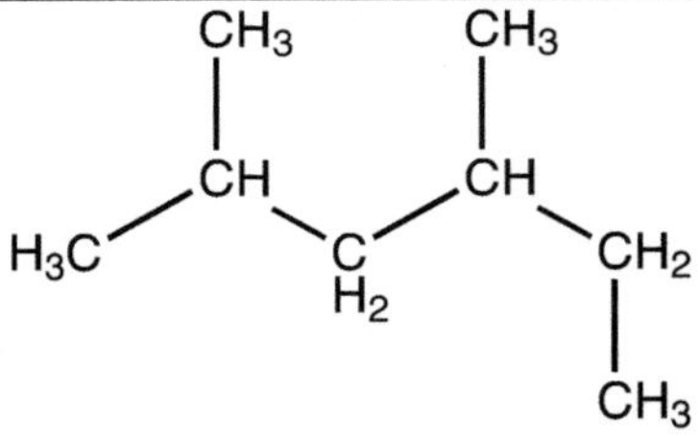

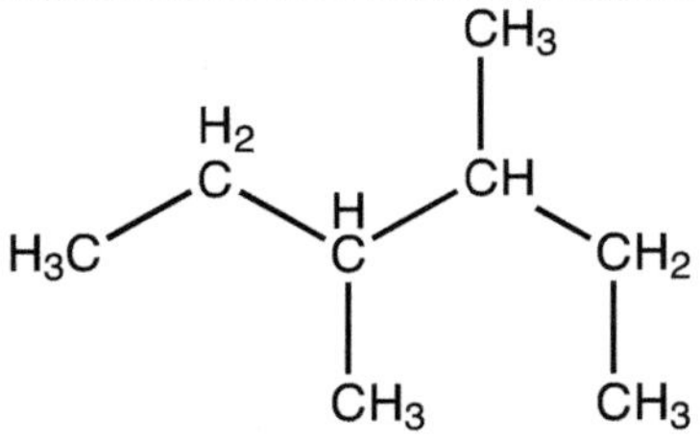

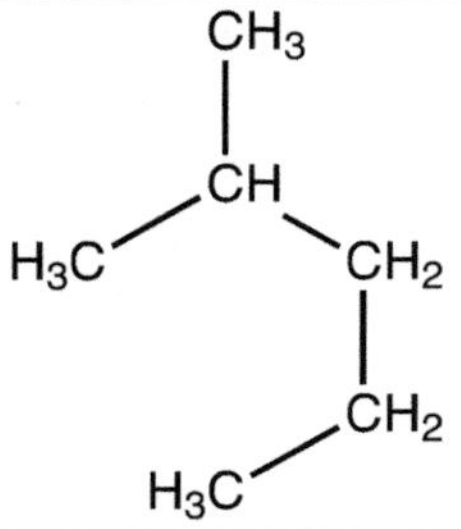

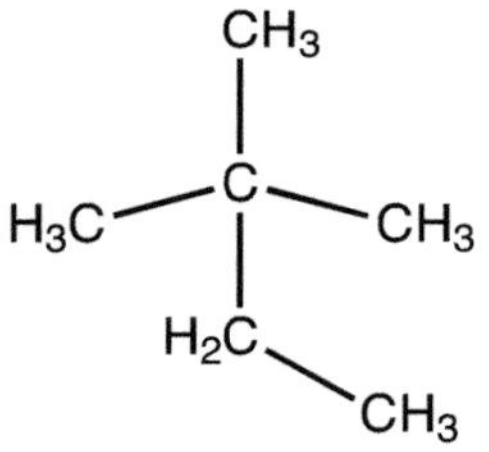

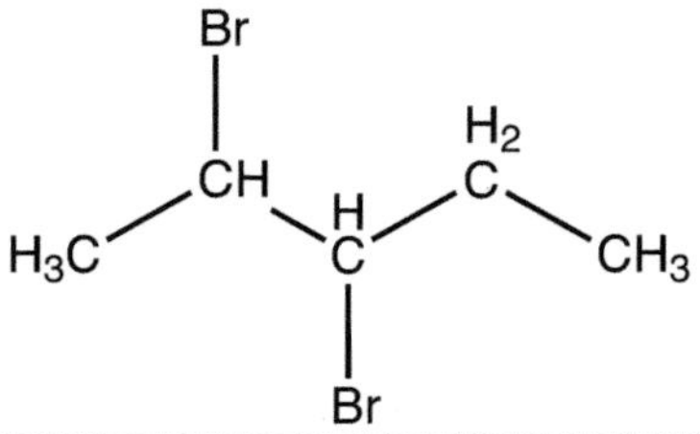

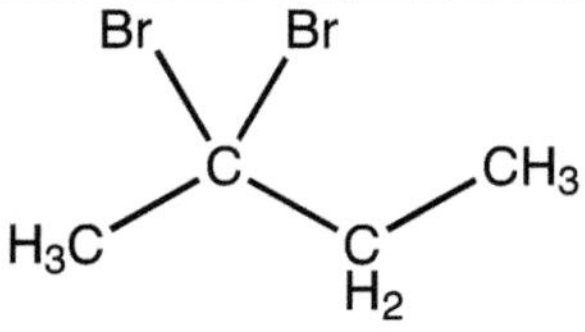

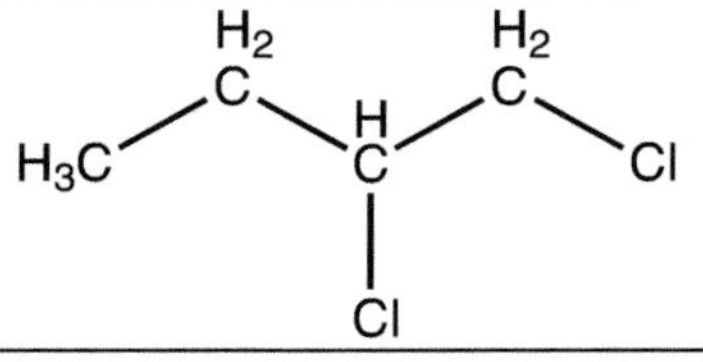

Isomers

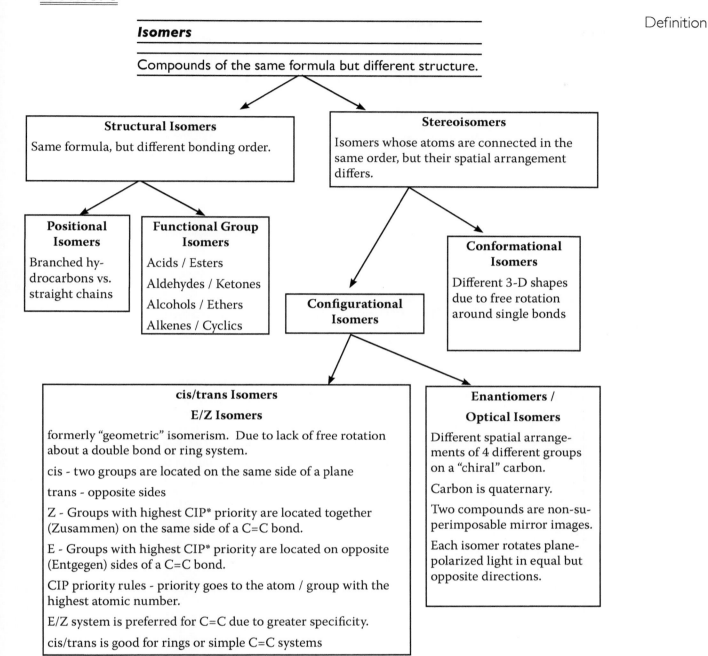

Definition

Isomers do not change back and forth into each other. Two different compounds may be isomers, but compounds do not spontaneously turn into other compounds. There is no "resonance" with isomers. "Moving" atoms around is only what we do on the page, not in the test tube!

Common Mistake

CIP Priority Rules

CIP Priority Rules state that we take the atom or group with the highest atomic number (or sum) as priority. So 1-fluoro-2-chloropropene can be "Z" when chlorine (17) and the methyl(15) groups are on the same side of the double bond, and "E", when on opposite sides.

Structural Isomers

Structural isomers are compounds that have the same formula but a different pattern of bonding. There are two sub-types of structural isomerism - Positional Isomerism and Functional Group Isomerism.

Positional isomerism probably the first type that you were taught. It's the difference between straight chains and branched chains.

butane (n-butane)	2-methylpropane
$H_3C-CH_2-CH_2-CH_3$	$(CH_3)_2CH-CH_3$

Table 10.6: The two isomers of C_4H_{10}

10.2 Learning Check

Name and draw the 3 isomers of pentane, C_5H_{12}
Name and draw the 5 isomers of hexane, C_6H_{14}

Branching in Organic Chemistry

When describing the branching in organic chemistry, we refer to the number of carbon atoms that are attached to the carbon atom of interest.

Description	bonding	comment
Primary	1 carbon neighbor	must be a CH_3 at the end of a chain
Secondary	2 carbon neighbors	the CH_2 carbons in the middle of the chain
Tertiary	3 carbon neighbors	a branching point in the chain
Quaternary	4 carbon neighbors	two branches at the same carbon in the chain

Figure 10.7: 2,2,3-trimethylpentane

2,2,3-trimethylpentane shows examples of each type of carbon. This can also be seen in nitrogen compounds - particularily amines

Primary amine	Secondary amine	Tertiary amine
H_3C-NH_2	$H_3C-NH-CH_3$	$H_3C-N(CH_3)-CH_3$
methylamine	diethylamine	triethylamine

Alkanes

Alkanes are the "backbone" of organic chemistry. They are generally unreactive. Their inert nature is due to the low electronegativity difference between carbon (2.4), and hydrogen (2.1). This makes the molecule non-polar. The carbon - hydrogen bond is also rather strong (412 kJ mol^{-1}). The strong bond is hard to break, so the chain often remains intact.

They provide a structure for chemists (and nature) to hang off more interesting reactive groups – functional groups.

There are however two types of reactions that alkanes undergo.

Combustion Reaction of Alkanes

All alkanes completely combust when reacted with excess $O_2(g)$ to produce $CO_2(g)$ and $H_2O(g)$.

Incomplete combustion occurs when their isn't sufficient oxygen, and produces either carbon monoxide or carbon solid (soot)

complete: $\quad C_5H_{12} + 8O_2(g) \rightarrow 5CO_2(g) + 6H_2O(g)$

incomplete $\quad C_5H_{12} + {}^{11}/_2 O_2(g) \rightarrow 5CO(g) + 6H_2O(g)$

incomplete $\quad C_5H_{12} + 3O_2(g) \rightarrow 5C(s) + 6H_2O(g)$

Carbon monoxide (CO), is a colourless, odourless, toxic gas that is carried more readily by the blood than oxygen.

Write the balanced equations for the complete combustion of the first four alkanes.

10.3 Learning Check

Write the balanced equations for the combustion of the first four alkanes producing carbon monoxide.

10.4 Learning Check

Substitution Reaction of Alkanes

Alkanes can react with halogens **in the presence of ultraviolet light** to exchange a hydrogen for a halogen

$$CH_3CH_3(g) + Br_2(g) \xrightarrow{U.V.\ light} CH_3CH_2Br(g) + HBr$$

The ultraviolet light is required to break the Br–Br bond **homolytically**, producing two free radicals: Br•. As opposed to **heterolytically** which would produce Br⁺ and Br⁻.

The reactants are orange / brown due to the $Br_2(g)$ present, the products are colourless.

Br_2 + ultraviolet light $\rightarrow$ 2Br•	**production** of bromine radical
CH_3CH_3 + Br• $\rightarrow$ CH_3CH_2• + HBr	**propagation** step
CH_3CH_2• + Br_2 $\rightarrow$ CH_3CH_2Br + Br•	**propagation** step
CH_3CH_2• + Br• $\rightarrow$ CH_3CH_2Br	**termination**: recombination of radicals

Substitution of more hydrogens can take place and eventually replace all H's with Br's. This reaction needs the ultraviolet light (a test tube placed in sunlight is sufficient) in order to react.

Organic Functional Groups

Functional Group	Type	Condensed Structural Formula	Structural Formula	4 carbon example
Hydrocarbons	Alkanes	R-CH$_3$	R—CH$_3$	Butane
Hydrocarbons	Alkenes	RCH=CH$_2$	R—CH=CH$_2$	2-Butene / But-2-ene
Hydrocarbons	Alkynes	RC≡CH	R—C≡CH	But-1-yne
Alcohols	Primary (1°)	R-CH$_2$OH	R—CH$_2$—OH	1-Butanol / Butan-1-ol
Alcohols	Secondary (2°)	R-CH(OH)R'	R—CH(OH)—R	2-Butanol / Butan-2-ol
Alcohols	Tertiary (3°)	R-CR'(OH)R''	R—CR(OH)—R	2-methylpropan-2-ol
Aldehydes	Aldehydes are always terminal	R-CHO	R—CHO	Butanal
Carboxylic Acids	Acids are always terminal	R-COOH	R—COOH	Butanoic Acid

Organic Chemistry

Functional Group	Type	Condensed Structural Formula	Structural Formula	4 carbon example
Esters	The acidic H has been replaced by a chain	R-COO-R'	R-C(=O)-O-R	Methyl Propanoate
Ketones	Ketones are always secondary	R-CO-R'	R-C(=O)-R	Butanone
Amines	May also be 2° or 3°	R-NH$_2$	R-CH$_2$-NH$_2$	Butamine
Nitriles	always terminal	R-CN	R-C≡N	Butanenitrile
Amides	Nitrogen analogue of carboxylic acids.	R-CO-NH$_2$	R-C(=O)-NH$_2$	Butamide

"Organic compounds are thought of as consisting of a relatively unreactive backbone, for example a chain of sp^3 hybridized carbon atoms, and one or several functional groups. The functional group is an atom, or a group of atoms that has similar chemical properties whenever it occurs in different compounds. It defines the characteristic physical and chemical properties of families of organic compounds." (IUPAC)

A functional group is the reactive part of the molecule. Is is so because the group imparts some polarity or bond type that can react. An alkane itself is generally unreactive as the C-C and C-H bonds are both strong and non-polar.

IB Chemistry is not generally concerned with multifunctional compounds with the exception of condensation polymerization monomers. Therefore, if asked to draw the possible structures for a given formula, ONLY USE THE ABOVE FUNCTIONAL GROUPS.

Students often draw "enols" as an isomer of $C_nH_{2n}O$. An enol contains a C=C and an OH in the same molecule. DO NOT USE THESE. (They do exist, but not in the IB Chemistry world.) If you draw one, you will get yourself on a path where the rest of the question stops making sense. $C_nH_{2n}O$ is either going to be an aldehyde or a ketone, and $C_nH_{2n}O_2$ will be an acid or ester.

Exam Trap

Functional Groups vs. Classes of compounds

In some cases a functional group is the same as the name of the class of compounds, in other cases a class of compounds can contain more functional groups

Functional Group	Structure	Classes containing Functional Group
hydroxy *	-OH	alcohols, acids
carbonyl	-C=O	aldehydes, ketones, amides
carboxyl	-C(=O)O-	carboxylic acids, esters
ether	-C-O-C-	ethers
amines	R-NH$_2$	amines (primary, secondary, tertiary)
nitriles	C≡N	nitriles

* Don't confuse with "hydroxyl", which indicates the OH• radical.

Functional Group Isomerism

Isomerism can also give rise to different functional groups. It arises from the interchange of atoms that are specific to a functional group.

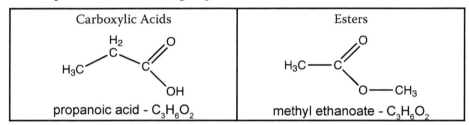

Figure 10.8: Acid & Ester Isomers

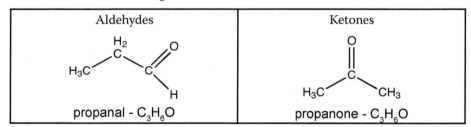

Table 10.9: Aldehyde & Ketone Isomers

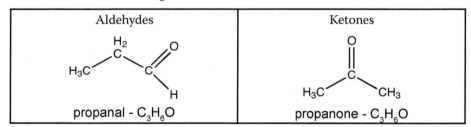

Figure 10.10: Alcohol & Ether Isomers

Figure 10.11: Alkene & Cyclic Isomers

Organic Chemistry

Draw, and name all non-branching isomers of...
a) $C_4H_8O_2$
b) $C_5H_{10}O$

10.5 Learning Check

Alkenes — Electrophilic Addition Reactions

Alkenes are a basic starting point for organic chemistry as they have a region of accessible attractive electron density. The electron density is accessible because it is in a π bond which has a high chance of collision with another species that is partially positive or electron deficient - an electrophile. An electrophile is a Lewis acid - it can accept a pair of electrons.

There are five addition reactions of alkenes. These are shown in the following 5 diagrams.

Figure 10.37: The pi bond in ethene

1. Hydrogenation (requires a catalyst of Ni, Pd, or Pt)

propene + H_2(g) →(catalyst) propane

2. Halogenation

propene + Br_2 → 1,2-dibromopropane

3. Addition of HX

2-bromopropane
major product

1-bromopropane
minor product

4. Hydration - Addition of water.

$H_3C-CH=CH_2 + H-OH$

with H_3PO_4 catalyst gives:

Major product: $CH_3-CH(OH)-CH_3$ — propan-2-ol

Minor product: $CH_3-CH_2-CH_2OH$ — propan-1-ol

5. Addition Polymerization

propene $\xrightarrow{poly}$ polypropene

$$\left(\begin{array}{c} H \\ | \\ -C- \\ | \\ CH_3 \end{array} \begin{array}{c} H_2 \\ | \\ -C- \end{array} \right)_n$$

Be careful with the drawing of the polymer. **The polymer chain is built of only the two carbons involving the double bond.** All other parts of the molecule are branches off of the chain. In the example of polypropene, there is a methyl group branching off of the main polymer chain every two carbons.

Below are some examples of addition monomers and their corresponding polymers.

Organic Chemistry

Monomer	Polymer repeating unit
 $H_2C=CH_2$ ethene	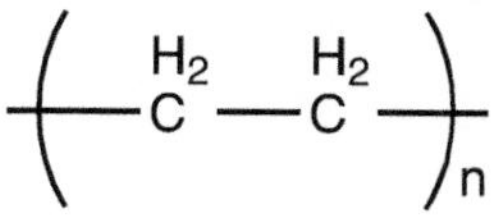 $-(-CH_2-CH_2-)_n-$ polyethene
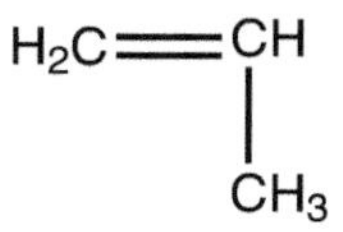 $H_2C=CH-CH_3$ propene	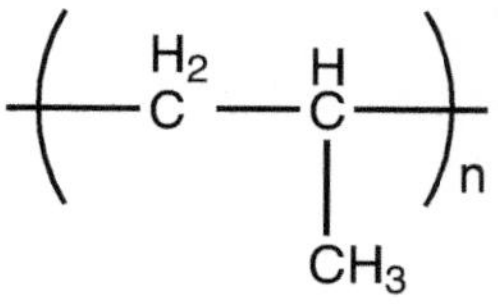 $-(-CH_2-CH(CH_3)-)_n-$ polypropene
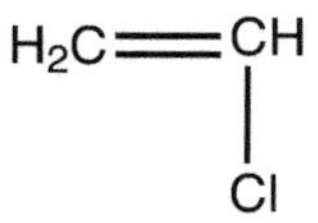 $H_2C=CH-Cl$ chloroethene (old name = vinylchoride)	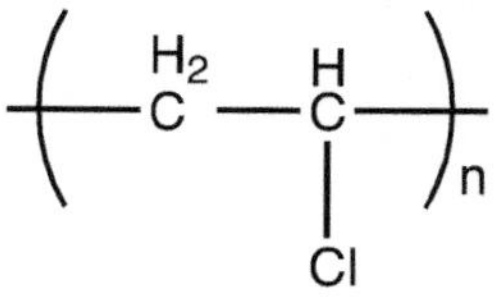 $-(-CH_2-CH(Cl)-)_n-$ polychloroethene (old name = polyvinylchloride - PVC)
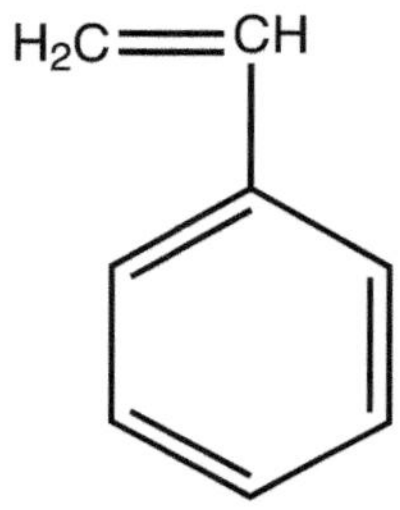 $H_2C=CH-C_6H_5$ styrene	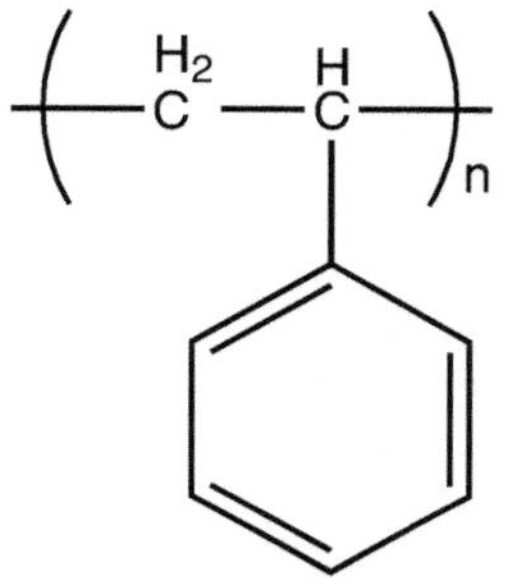 $-(-CH_2-CH(C_6H_5)-)_n-$ polystyrene

Table 10.12: Examples of monomers and polymers

Draw the polymer repeating unit for the following monomers.

a) but-2-ene
b) but-1-ene

10.6 Learning Check

Electrophilic Addition Mechanism

The electrophilic addition mechanism gives rise to two possible products. in 1870, Russian Chemist, Vladimir Markovnikov worked out the rule to determine which product was more likely. It turns out that the rule is based on the stability of the carbocation intermediate. We know that tertiary carbocations are more stable than secondary, than primary carbocations, and this gives rise to the explanaiton of which product will dominate.

In the mechanism, the polar HBr molecule aligns itself with the more positive H being attracted to the C=C double bond, an electron rich bond. The pi bond "reaches out" to the hydrogen. The C-H bond can form from either carbon, but the location of the carbocation is the thing to notice.

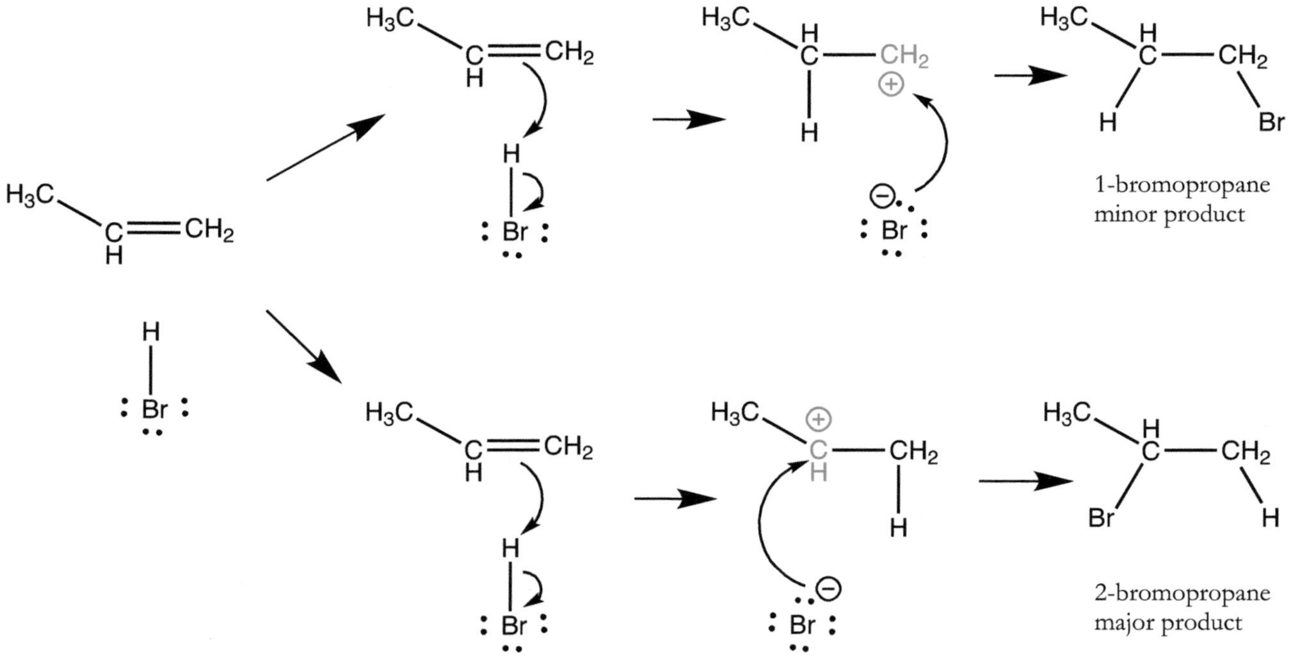

1-bromopropane
minor product

2-bromopropane
major product

A quick way to remember what goes where is that the hydrogen atom (of H-X) will go where there are already the most number of hydrogens. In the major product above you can see that the hydrogen adds to the CH_2 group, not the CH group.

Benzene and Derivatives

As mentioned in the Bonding unit, there is a ring structure containing 6 carbons and 6 hydrogens. The hexagonal ring consists of alternating single and double bonds which gives rise to a resonance delocalization.

Benzene does not undergo the electrophilic addition reaction above.

Benzene itself is C_6H_6 and the C_6H_5 group is called "phenyl" (like "methyl"). As compounds containing phenyl groups are known as aromatic compounds, these groups are also called "arenes".

Although benzene appears to have double bonds in it, it does not behave like an alkene due to the delocalization.

Organic Chemistry

Important isomers

You should be familiar with the four isomers of substituted butanes, as they represent structural isomers as well as the concept of positional isomers and compounds containing primary, secondary and tertiary carbons.

The degree of substitution of a carbon indicates the number of branches that are attached to it. Primary carbons have only one branch - and are therefore at the end of a chain. Secondary carbons are in the middle of a chain. Tertiary carbons have three carbon chain branches. We usually identify the degree of a carbon bearing a halogen or an alcohol.

1-bromobutane (1°/primary)	2-bromobutane (2°/secondary)
1-bromo-2-methylpropane (1°/primary)	2-bromo-2-methylpropane (3°/tertiary)

Table 10.13: Isomers of C_4H_9Br

Be able to draw these structures

1-butanol / butan-1-ol (1°/primary)	2-butanol / butan-2-ol (2°/secondary)
2-methylpropan-1-ol (1°/primary)	2-methylpropan-2-ol (3°/tertiary)

Table 10.14: Isomers of C_4H_9OH

Be able to draw these structures

Often candidates are asked to draw either of the four sets above. Make sure you know them.

Halogenoalkanes

Some halogenoalkanes are generally not very reactive. They get used as a variety of solvents which can end up in the upper atmosphere and cause ozone damage.

They do however provide a small amount of reactivity because they introduce a region of polarity into the molecule. The carbon – halogen bond is polar due to the high electronegativity of the halogen. This polarity allows the carbon atom to be partially positive (∂+) and therefore be susceptible to attack by something that is attracted to a partially positive charge – a nucleophile.

10.7 Learning Check

Draw and name all 16 isomers of chlorohexane, $C_6H_{13}Cl$. Identify each as primary, secondary or tertiary with regards to the halogen bearing carbon.

Nucleophilic Substitution reactions

In a nucleophilic substitution reaction, a "nucleophile" replaces a "leaving group".

Nucleophile: an ion or molecule with at least one lone pair of electrons which, is attracted to a partial positive charge centre.

(NOTE: not attracted to "a nucleus")

In general a hydroxide ion displaces a halogen from a molecule.

$$OH^- + R\text{—}X \rightarrow R\text{—}OH + X^-$$

There are two different mechanisms through which this substitution can occur. One uses first order kinetics, and is termed S_N1, the other uses second order kinetics and is termed S_N2.

S_N1 – Nucleophilic Substitution – First order kinetics

The steps in the mechanism are as follows...
1. Dissociation of the leaving group (the halogen) to leave an ionic intermediate. This is the slow step. Therefore rate = $k[CR_3X]$
2. Attack by a nucleophile eg. OH^-, CN^-, H_2O or NH_3.

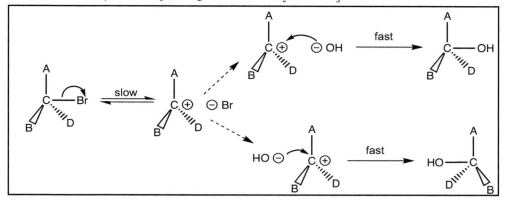

Figure 10.15: S_N1 reaction mechanism

*Exam Hint
Be able to draw the entire mechanism shown here*

Important details
1. The rate of reaction depends only upon the dissociation of the leaving group - this is the slow step.
2. The intermediate is ionic – it is stabilized by protic solvents (like water).
3. The intermediate is a carbocation (positive charge).
4. The intermediate is sp^2 hybridized and is planar triangular in shape.
5. The nucleophilic attack has a 50/50 chance of occurring on either side.
6. There are two optical isomers formed in equal quantities forming a *RACEMIC* mixture.
7. The optical isomer is centered around a quaternary carbon - called an asymmetric carbon.

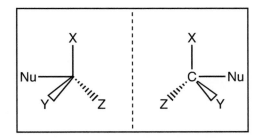

Figure 10.16: The two mirror images formed by an S_N1 reaction

A racemic mixture is a mixture of equal quantities of two optical isomers whose optical activities cancel out.

S_N2 – Nucleophilic Substitution – Second order kinetics

The steps in the mechanism are as follows...
1. Attack by the nucleophile on the substrate
2. Dissociation of the leaving group.

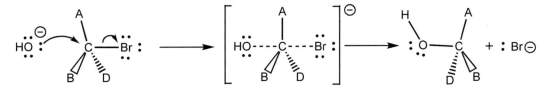

Figure 10.17: Sn2 Reaction Mechanism

Important details
1. The rate of reaction depends on the concentration of both the nucleophile and the substrate. Rate=k[CR_3X][Nu]
2. The intermediate is non-ionic, and is stabilized by aprotic solvents, like propanone.
3. The intermediate has three full strength bonds, and two partial bonds.
4. As the charge density (bond) between the nucleophile and the substrate increases, the charge density between the substrate and the leaving group decreases. – one fades in the other fades out.
5. The intermediate is a trigonal bipyramidal shape.
6. The nucleophile must attack from the side opposite the leaving group.
7. There is only one optical isomer product, which has inverted stereochemistry from the reactant.

Which Mechanism? S_N1 or S_N2 – What's the difference?

Factor	S_N1	S_N2
Solvent	protic - water	aprotic – like propanone
Structure	3°/ Tertiary carbons	1°/ Primary carbons
Branching	Bulky / branched chains prevent backside attack. This is called "stearic hindrance".	Linear side chains or hydrogen allow space for nucleophile to attack first.
Bond strength of leaving group	High	low

What about 2° carbons? – You have to look at the other factors involved – they will determine the dominant mechanism.

Nucleophilic Substitution Reactions II

There are a variety of factors that affect rate and products of nucleophilic substitutions. They are

1. Identity of the halogen (leaving group)
2. Structure of the carbon skeleton of the halogenoalkane
3. Identity of nucleophile

Identity of the halogen (F, Cl, Br, I)

If we compare the reactivity of halogenoalkanes by using different halogens, we find that the rates of reaction increase as we move down group 7. This is due to the weakening of the C-X bond. (not electronegativity)

Check the "Average Bond Enthalpies" in your data booklet to confirm and write the values here...

Bond	Bond Enthalpy
C-F	
C-Cl	
C-Br	
C-I	

fast----slow

Structure of the carbon skeleton

In an S_N2 reaction, the nucleophile must be able to reach the $\partial+$ carbon from the side opposite the leaving group. If there are side chains in the way, then there is no space for the nucleophile to attack. This is called "stearic hindrance". Stearic hindrance is greatest in bulky side chains and lowest in primary carbons, which only bear hydrogens.

Substrate	side chains	mechanism
Primary	2 hydrogen atoms, main chain	S_N2
Secondary	1 hydrogen, two alkyl branches	S_N1 or S_N2
Tertiary	0 hydrogens, 3 alkyl branches - linear	S_N1

Identity of Nucleophile

There are lots of nucleophiles that are available to produce a variety of products.

Nucleophile	Product
OH⁻	alcohol
CN⁻	nitrile
NH_3	amine
H_2O	alcohol

Table 10.18: Other nucleophiles

Alcohols

Alcohols are the homologous series that contain an OH group.

The difference between NaOH and alcohols is that NaOH is ionic and dissociates into Na⁺(aq) and OH⁻ (aq) ions, whereas C–OH is covalent and does not dissociate – It's a molecule.

Short chain alcohols are miscible with water (they can mix in any proportion), due to the hydrogen bonding of the OH group. As the chain lengthens, the solubility decreases because the proportion of hydrogen bonding in the molecule is overwhelmed by the London forces.

n	name	molecular formula	condensed structural formula	b.p	Solubility
1	methanol	CH_3OH	CH_3OH	65°C	miscible
2	ethanol	C_2H_5OH	$CH_3\text{-}CH_2\text{-}OH$	78°C	miscible
3	propan-1-ol	C_3H_7OH	$CH_3\text{-}CH_2\text{-}CH_2\text{-}OH$	97°C	miscible
4	butan-1-ol	C_4H_9OH	$CH_3\text{-}CH_2\text{-}CH_2\text{-}CH_2\text{-}OH$	118°C	9.1 g /100 g H_2O
5	pentan-1-ol	$C_5H_{11}OH$	$CH_3\text{-}CH_2\text{-}CH_2\text{-}CH_2\text{-}CH_2\text{-}OH$	138°C	2.7 g /100 g H_2O
6	hexan-1-ol	$C_6H_{13}OH$	$CH_3\text{-}CH_2\text{-}CH_2\text{-}CH_2\text{-}CH_2\text{-}CH_2\text{-}OH$	151°C	slightly soluble

Table 10.19: Homologous Series of Alcohols

Combustion of alcohols

Alcohols can be combusted - usually completely.

When balancing, don't forget about the oxygen in the alcohol.

10.8 Learning Check

Write balanced equations for the complete combustion of the first four alcohols.

Oxidation of alcohols

Alcohols may be oxidized by a variety of agents. The common one in IB Chemistry is "acidified potassium dichromate" or if you prefer $H^+/K_2Cr_2O_7$. (It needs to be acidic for it to work!)

Depending on the degree of your alcohol, you get different products, or none at all.

Primary alcohols can be oxidized to aldehydes, which can then be further oxidized to carboxylic acids. In all cases, the functional groups are terminal (at the end of the chain)

Secondary alcohols have an alcohol group in the middle of the chain, and can be oxidized to form ketones, which cannot be further oxidized.

Tertiary alcohols cannot be oxidized.

	Ox. agent	1st Product	Ox agent	2nd Product
1°	$H^+(aq)/Cr_2O_7^{2-}$ (aq)	aldehyde	$H^+(aq)/Cr_2O_7^{2-}$ (aq)	carboxylic acid
2°	$H^+(aq)/Cr_2O_7^{2-}$ (aq)	ketone	$H^+(aq)/Cr_2O_7^{2-}$ (aq)	none
3°	$H^+(aq)/Cr_2O_7^{2-}$ (aq)	none		

What you see during the oxidation is the oxidizing agent – the orange $Cr_2O_7^{2-}$ ion changing to the green Cr^{3+} ion (as the dichromate ion is reduced)

Remember to state the <u>change of colour</u> – *"orange to green"*. This is also a way to chemically distinguish tertiary alcohols from primary or secondary.

Alcohol Oxidation Products

When a primary alcohol is oxidized, we can control the type of product formed by the reaction conditions.

Consider the fact that alcohols have an OH bond, and thus, hydrogen bonding, whereas aldehydes have weaker dipole-dipole forces. Therefore the boiling point of the aldehyde is lower.

As soon as the lower boiling point aldehyde is formed, it instantly vaporizes and escapes the reaction mixture.

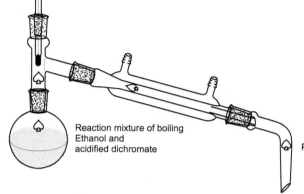

Reaction mixture of boiling Ethanol and acidified dichromate

Pure ethanal distilled off

If we distill of this vapour we can have a product of only the aldehyde.

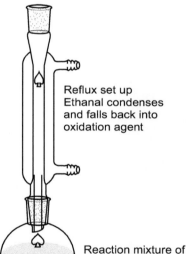

Reflux set up
Ethanal condenses and falls back into oxidation agent

Reaction mixture of boiling ethanol and acidified dichromate

If however, we need to produce the carboxylic acid product, the intermediate product, ethanal, must be oxidized a second time. We can do this simply by setting the apparatus to "reflux". The ethanal vapour is condensed in a vertical condenser and the liquid falls back into the reaction mixture to be further oxidized.

When the reaction is complete, we change the apparatus to distil the carboxylic acid product out of the reaction mixture.

In both processes, the oxidizing agent changes colour from orange to green.

Ethanol	Ethanal	Ethanoic acid
hydrogen bonding	dipole forces	strong hydrogen bonding
b.p. = 78°C	b.p. = 20°C	b.p. = 118°C

Reduction of aldehydes, ketones and acids

All of the oxidation products can be reduced back to their starting alcohols by lithium aluminium hydride, LiAlH4, a strong reducing agent.

	Reduced		oxidized
Primary	butan-1-ol	$\xrightleftharpoons[\text{LiAlH}_4]{H^+ / Cr_2O_7^{2-}}$	butanal
Secondary	butan-2-ol	$\xrightleftharpoons[\text{LiAlH}_4]{H^+ / Cr_2O_7^{2-}}$	butanone

Functional Groups containing Nitrogen

There are three nitrogen containing functional groups for you to know - Amines, Amides, and Nitriles

Amines are organic derivatives of ammonia. One, two or three of the hydrogens may be replaced by alkane chains. Amines are weak bases. Short chain alkanes are soluble in water due to hydrogen bonding.

ammonia	ethylamine

Figure 10.20: Ammonia and ethylamine

Other than their alkaline properties, amines can undergo condensation reactions with carboxylic acids to produce "amides", and a water molecule.

Amides are the nitrogen analogues of esters.

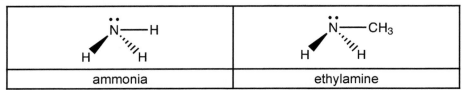

Ester RCOOR'	Amides RCONHR'	

Figure 10.21: Comparison of an ester and two amides

Nitriles contain a carbon-nitrogen triple bond. The formation of nitriles is important in that the addition of C≡N allows the carbon chain to be lengthened by one carbon. The cyanide ion CN⁻ is a strong nucleophile and can react by an S_N2 mechanism.

H-C≡N	CH₃C≡N
Hydrogen cyanide	ethanenitrile

Figure 10.22: Hydrogen cyanide and ethanenitrile

Organic Chemistry

The IUPAC nomenclature rules for naming nitriles states that *"in systematic nomenclature, the suffix nitrile denotes the triply bound ≡N atom, not the carbon atom attached to it."*

Carbons	Amines	Amides	Nitriles
1	methylamine	methanamide	hydrogen cyanide
2	ethylamine	ethanamide	ethanenitrile
3	propylamine	propanamide	propanenitrile
4	butylamine	butanamide	butanenitrile
5	pentylamine	pentanamide	pentanenitrile
6	hexylamine	hexanamide	hexanenitrile

Table 10.23: Names of the first six nitrogen containing compounds

Both ammonia and the cyanide ions can react with halogenoalkanes by a nucleophilic substitution to produce nitrogen containing products.

Figure 10.24: S_N2 reaction with ammonia forming an amine

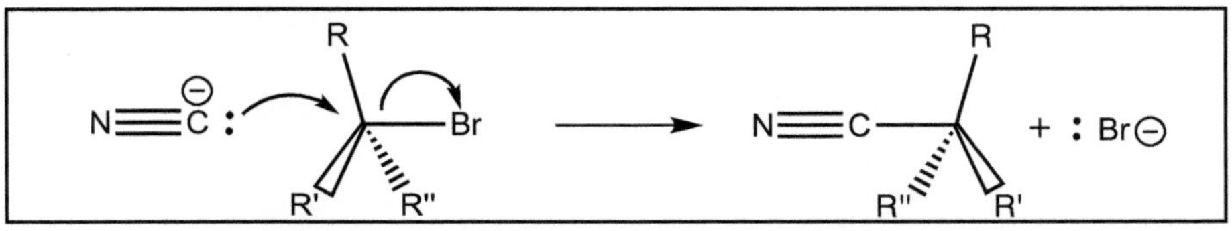

Figure 10.25: S_N2 reaction forming a nitrile

Reduction of Nitriles

Nitriles can be reduced with hydrogen gas and a nickel catalyst to form the amine. This is different from forming the amine directly by substitution, as we have lengthened the chain by one carbon. Why would we bother using the two steps of forming the nitrile, and then reduction to the amine, when we can form the amine directly? This way allows us to lengthen the chain by one carbon, and the cyanide ion is a stronger nucleophile than ammonia.

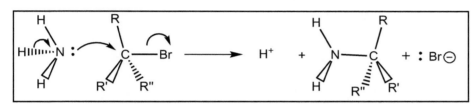

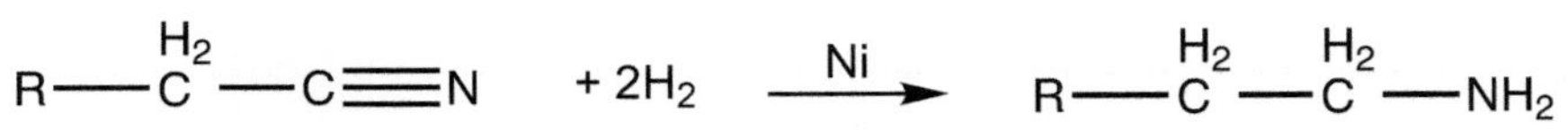

Figure 10.26: Reduction of a nitrile with hydrogen and a nickel catalyst

Elimination Reactions

The general reaction is the elimination of a proton and a leaving group forming a by-product such as HBr, and the organic compound forms a double bond.

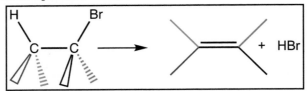

Figure 10.27: General Elimination Reaction

There are two mechanisms for elimination, E1 and E2. These correspond to the S_N1 and S_N2 mechanisms in terms of kinetics and elementary steps.

The mechanism for most elimination reactions is a one step mechanism which reminds us of the S_N2 mechanism, so it is called "E2".

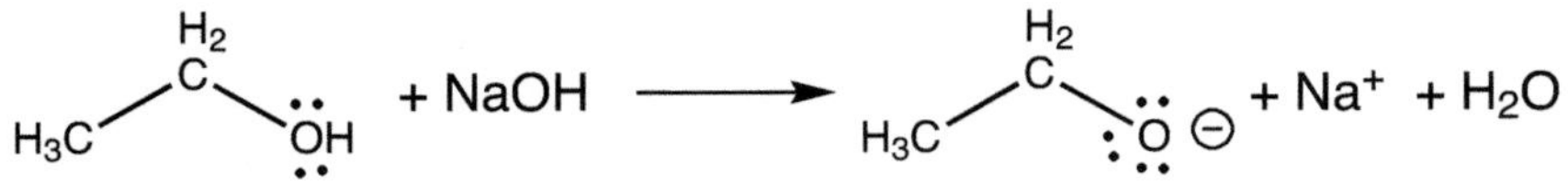

$$CH_3CH_2OH + NaOH \longrightarrow CH_3CH_2O^- + Na^+ + H_2O$$

The NaOH takes a proton from the ethanol to form the ethoxide ion.
This is not part of the reaction kinetics, but the conditions.
This is why you need "hot alcoholic sodium hydroxide".

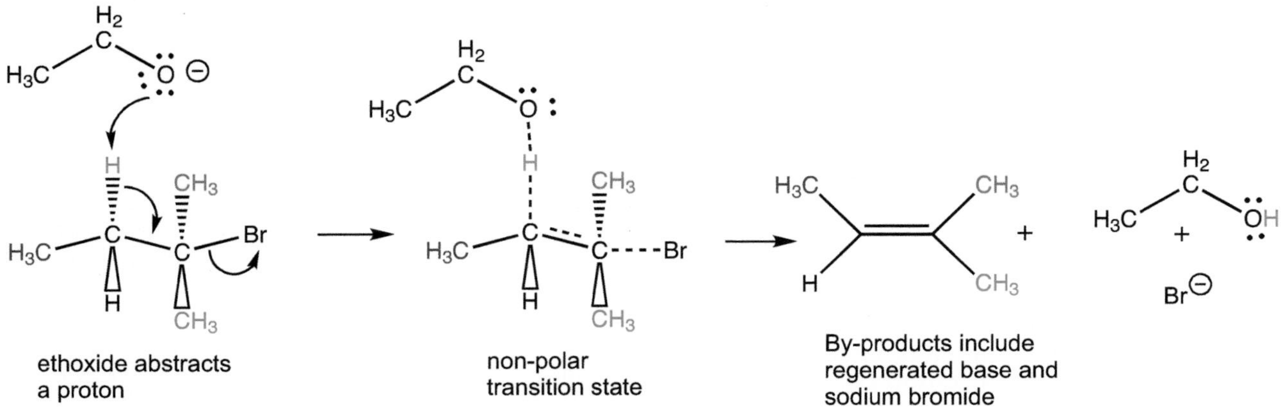

Figure 10.28: E2 Mechanism

The other elimination reaction is the E1 mechanism, which starts with a carbocation, and therefore is a competitor with the S_N1 reaction.

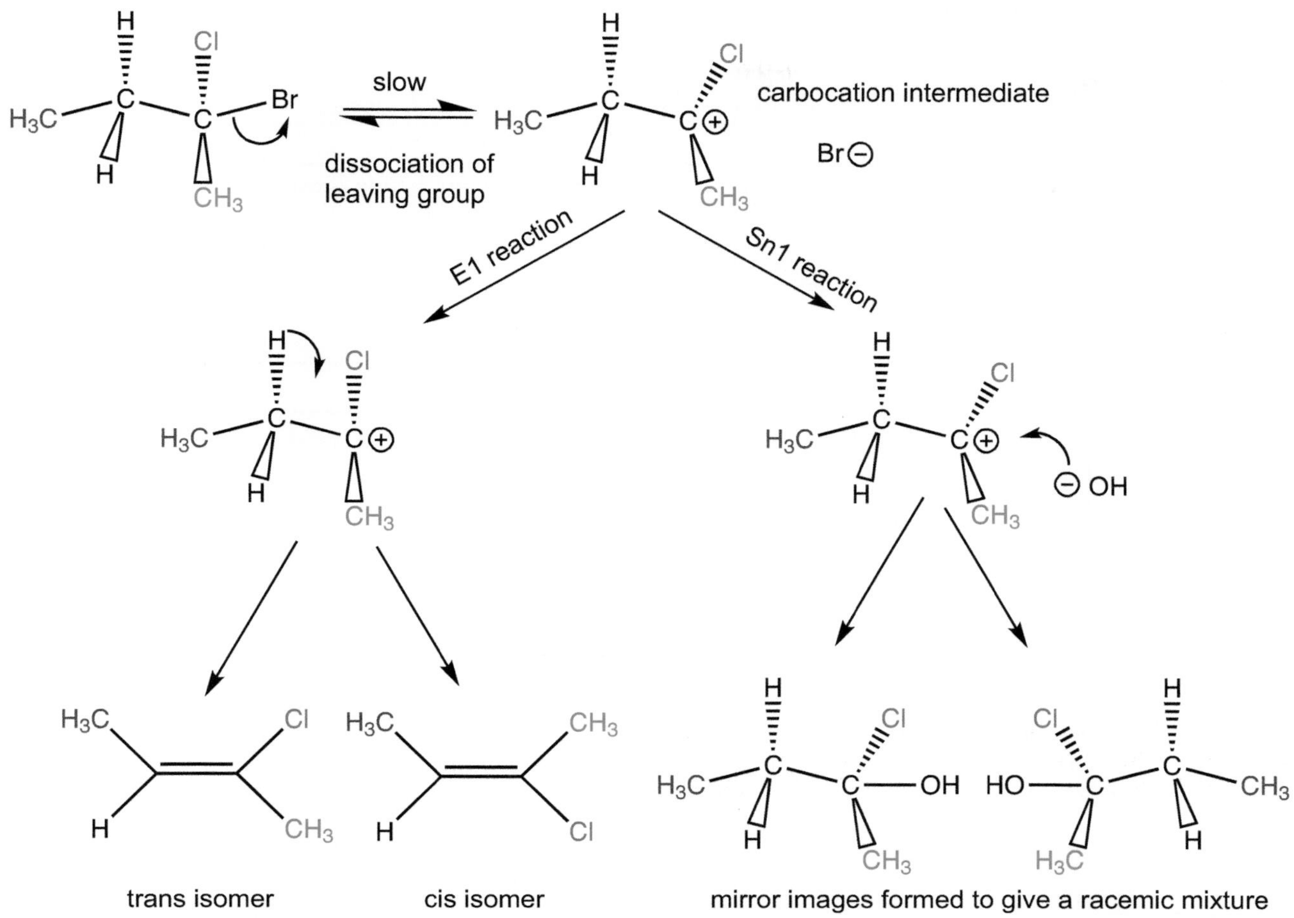

Organic Chemistry

Figure 10.29: E1 Mechanism

(trans isomer, cis isomer, mirror images formed to give a racemic mixture)

Ester Condensation Reactions

Condensation reactions are those which produce a small molecule, like H_2O, when a new bond is formed between two molecules. - This is not called a "dehydration" reaction - dehydration occurs when one molecule loses the equivalent of water.

There are two main organic products formed by condensation reactions in IB Chemistry, esters and amides.

Esterification is catalysed by acid. It is useful to use concentrated sulphuric acid for the catalyst because sulphuric acid has a high affinity for water, and thus drives the equilibrium to the right.

An ester is the result of the condensation between a carboxylic acid and an alcohol. It may be written as

$$\text{R-COOH} + \text{HO-CH}_2\text{-R'} \xrightarrow{H^+} \text{R-COO-CH}_2\text{-R'} + H_2O$$

or using structures...

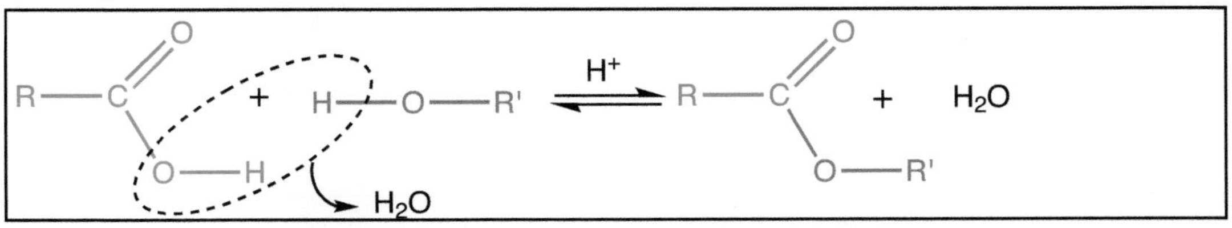

Figure 10.30: General Condensation Reaction

Figure 10.31: Condensation to form ethyl propanoate ester

Naming Esters

Esters are named by recognizing that the alcohol now forms a "side chain" off of the anion of the weak acid.

Reactants			Products		Ester name
Carboxylic Acid	Alcohol		Carboxylate ion	Side chain	
methanoic acid	ethanol		methanoate	ethyl	ethyl methanoate
ethanoic acid	methanol		ethanoate	methyl	methyl ethanoate
propanoic acid	butanol		propanoate	butyl	butyl propanoate
benzoic acid	propanol		benzoate	propyl	propyl benzoate

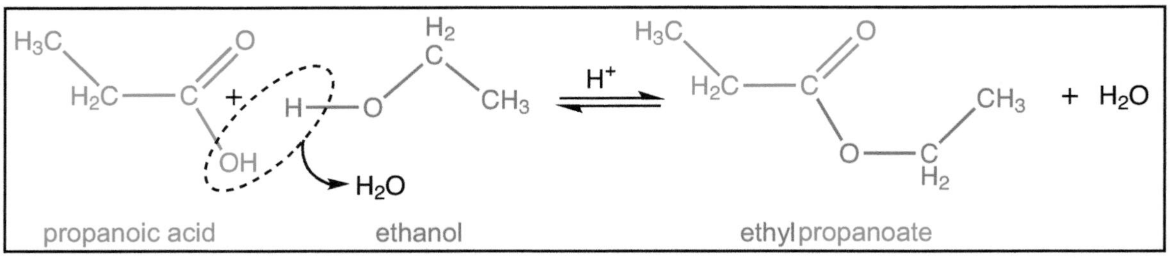

Remember that esters are functional group isomers of carboxylic acids. You can also write isomeric forms by changing around the relative lengths of the two chains.

For example "methyl ethanoate" and "ethyl methanoate" are isomers.

Most esters smell pleasant or fruity. Due to the lack of hydrogen bonding, they don't mix well with water and have relatively low boiling points, so you can smell them easily.

Amide Condensation Reactions

Amides are formed when the carboxylic acid reacts with an amine or ammonia.

$$R\text{-}COOH + NH_3 \rightleftharpoons RCOONH_2 + H_2O$$

$$R\text{-}COOH + NH_2\text{-}R' \rightleftharpoons R\text{-}CONHR' + H_2O$$

We cannot use acid catalysts for amide formation because the acid protonates the amine and prevents the lone pair from being able to attack the carbon of the acid. It is sufficient to warm the reaction mixture.

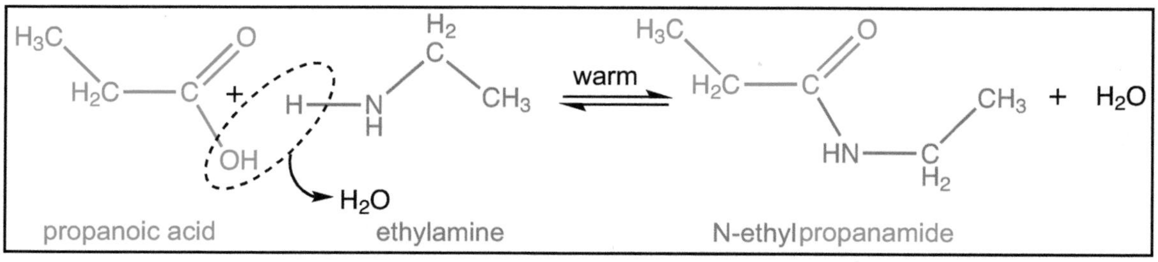

Figure 10.32: Condensation reaction to form N-ethylpropanamide

Organic Chemistry

Electrophilic Substitution Reactions

As previously discussed, benzene does not undergo electrophilic addition reactions, but the pi cloud is attractive to electrophiles, and can undergo substitution reactions.

The classic example of electrophilic substitution reactions is the nitration of benzene. This is carried out with "mixed acids" - nitric and sulfuric.

In the first step the two acids react to produce the "nitronium ion", NO_2^+ which is electron deficient, and therefore an electrophile.

$$HNO_3 + H_2SO_4 \rightarrow NO_2^+ + H_2O + HSO_4^-$$

This may be understood by writing the nitric acid as $HO-NO_2$ and the sulfuric acid in its dissociated form

$$HO-NO_2 + H^+ + HSO_4^- \rightarrow H_2O + NO_2^+ + HSO_4^-$$

In the second step the pi cloud electrons reach out to the electrophile, NO_2^+, disrupting the delocalization temporarily. The electrons "roll" around the ring and reform the delocalized pi ring.

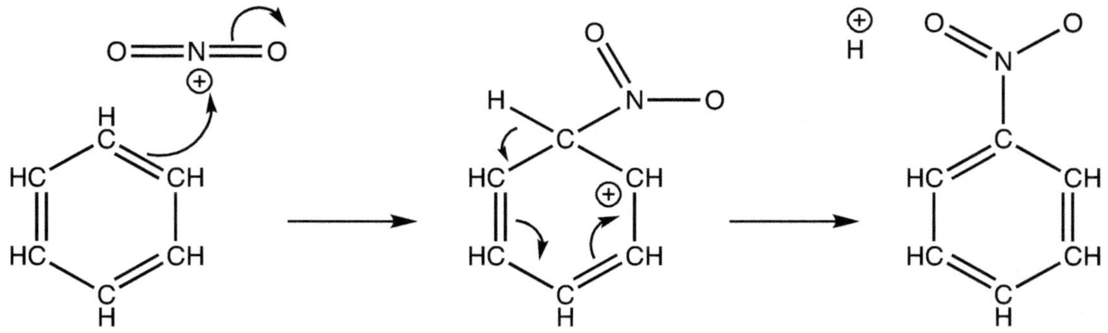

Figure 10.33: Nitration of Benzene

Reaction Pathways

Reaction pathways refer to the combination of reactions that lead to a desired product in four steps from a specific starting material.

The key is that you need to know all the products and reactants so that you can determine a reasonable intermediate product.

You may be required to state...

- the conditions for the first reaction.
- the isolation / collection of the intermediate product.
- the name and structure of the intermediate product.
- the type of intermolecular forces for the intermediate product.
- the conditions for the subsequent reaction(s).
- the isolation / collection of the final product.
- the type of intermolecular forces for the final product.
- the relative boiling points of each of the three species.

State the reactions and conditions needed to produce butan-2-one from but-2-ene.

State the reactions and conditions needed to produce propanal from 1-bromopropane.

10.9 Learning Check

---

1. State the reactions and conditions needed to produce propanenitrile from ethene.
2. State the reactions and conditions needed to produce propylamine from propene.

10.10 Learning Check

Stereoisomerism

Stereoisomers have the same bonding arrangement (the same atoms are bonded together), but their spatial arrangement is different. This occurs in two different situations.

E-Z isomers – Alkenes

E-Z isomerism arises from the fact that the pi bond does not allow for different conformations due to free rotation. Put simply there is no free rotation around the double bond.

In some cases determining which two parts of the molecule to look at are difficult, so three chemists invented the CIP(Cahn-Ingold-Prelog) priority rules. Simply put, the atom or group with the highest atomic number (or sum) gets priority. This creates an "absolute" reference, which is convenient as it is not always clear which reference plane to use.

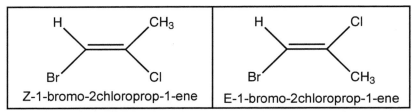

Figure 10.34: E-Z isomerism in 1-bromo-2-chloroprop-1-ene

Looking at the 4 groups we have Hydrogen (1), methyl (12+3=15), Chlorine (17) and bromine (35) arranged around the double bond. In the Z isomer the two highest priority groups, chlorine and bromine, are on the same side of the double bond - this is given the "Z" prefix for "zusammen" (German for "together"). In the E isomer, the two highest priority groups are on opposite sides of the double bond and are given the "E" prefix for "entgegen" (German for opposite).

Cis/Trans isomerism in rings.

In cyclic / ring systems, the ring defines the reference plane and therefore we can use the older cis/trans method. If the two substituents are on the same side of the ring they are given the "cis" prefix. If they are on opposite sides, they are said to be "trans"

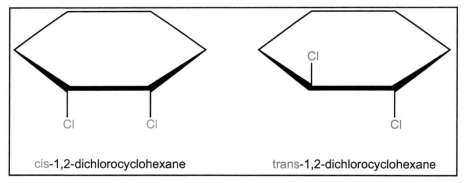

Figure 10.35: Geometric isomers of 1,2-dichlorocyclohexane

The two geometric isomers will have slightly different properties.

For example, due to electronegative groups on the same side of the double bond, Z-1,2-dichlorobut-2-ene has greater polarity than the E isomer. This gives the Z isomer a higher boiling point of 125°C, while the E isomer boils at 101°C.

Optical Isomerism

Optical isomers are called "enantiomers" (en-ant-ee-oh-mers). These are the mirror images that are typically formed in S_N1 reactions. Optical isomers have the property of optical activity.

Optical Activity: the ability of a substance to **rotate** plane polarized light.

Be careful to say "rotate" or twist", do not say "bend" - that's refraction.

Optical activity and optical isomerism arise from the molecule containing one or more "chiral" or "asymmetric" carbon atoms.

Chiral or Asymmetric carbon: A carbon atom with four different groups attached to it. Usually indicated in a molecule by an asterisk *.

To illustrate the point, glucose has four chiral carbons.

You must look at the entire part of the molecule around each carbon, not just the next atom.

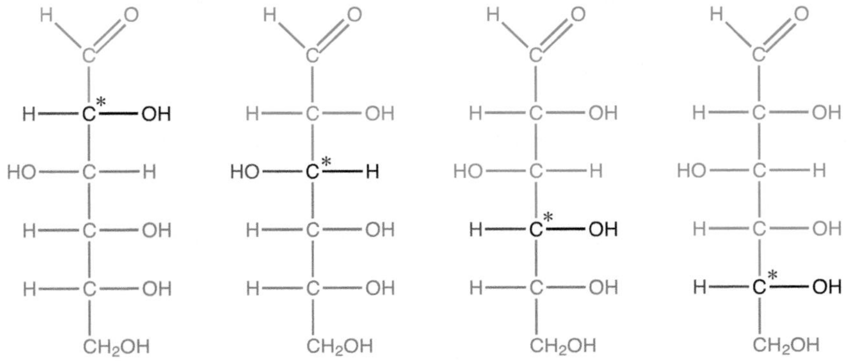

Figure 10.36: Four representation of D-Glucose, showing each chiral centre

In glucose, C-1, the top carbon atom in figure 10.25 is not chiral because it has only three different groups attached to it. The bottom carbon - C-6, is also not chiral because it has two identical hydrogen atoms attached to it.

If both optical isomers are present then the mixture is called "racemic"

A racemic mixture contains equal amounts of both optical isomers, and therefore the rotations cancel out.

Summary Questions

1. The following names are incorrect. Draw the structure and correctly name it.
   a) 2-ethylpropane
   b) 1-methylbutane
   c) 3-methylbutane
   d) 2,3-diethylbutane

2. Compound P has the formula $C_4H_{10}O$ and could exist as 4 different isomers.

   a) Name and draw all four isomers of P.

   b) i) One of the isomers of P may be oxidized to produce a non-acidic compound. Draw the molecule and name it.
   ii) State the conditions for this reaction.

   c) Another isomer of P cannot be oxidized. Name this isomer.

   d) Two of the isomers of P may be oxidized to produce carboxylic acids.
   i) What conditions are used to produce the acids?
   ii) Name and draw the two acid isomers.

   e) Compound P can be formed by two different types of nucleophilic substitution mechanisms.
   i) What is meant by the term "nucleophilic substitution"
   ii) Write a chemical equation for this reaction.
   iii) Determine the isomer that was formed by an S_N1 reaction. Choose an appropriate starting material (Q) and draw the mechanism.
   iv) Identify the isomers of P that were formed by an S_N2 reaction. Draw and name the starting materials.

   f) Compound Q can undergo an elimination reaction. Draw the mechanism for the reaction, draw and name the product.

   g) Identify the isomer of P that is optically active.

Chapter 11

Measurement & Data Processing

In this chapter...

- 182 Observations
- 182 Uncertainty in Measurements
- 182 Precision & Accuracy
- 183 Significant Digits
- 183 Rounding off Numbers
- 184 Adding and Subtracting Significant Digits
- 184 Multiplying and Dividing Significant Digits
- 184 Systematic Uncertainty
- 185 Absolute & Relative Uncertainty
- 185 Propagating Uncertainty
- 186 Types of Relationships
- 187 The Formula of a compound
- 188 Mass Spectroscopy (MS)
- 188 Nuclear Magnetic Resonance
- 189 Infrared Spectroscopy
- 189 X-ray Crystallography

Observations

Scientists make two kinds of observations, qualitative and quantitative. The former involves the non-numerical qualities, such as colour; whereas the latter are numerical measurements.

Uncertainty in Measurements

Unlike in Mathematics, numbers in Science are measurements. They result from the comparison of the magnitude of some quality against a standard. The quality of the measuring device tells us how precise our measurement can be and what the uncertainty is. Random uncertainty is associated with the measuring device and things like reaction time. Just because your stopwatch measures to 0.01 s, doesn't mean you do!

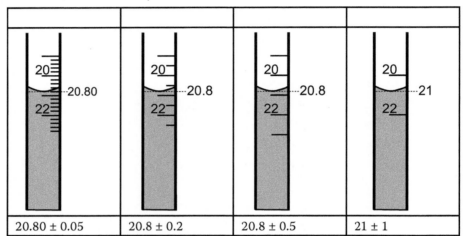

Figure 11.1: Volume measurements and uncertainty using different scales

Precision & Accuracy

Accuracy is the degree of exactness that the measurement has to the "true" value.

Precision refers to the degree of exactness which a measuring instrument can determine accuracy.

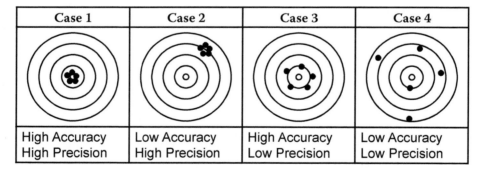

In case 1 - all shots hit very close to the center; all measurements are close to the true value, and each other.

In case 2 - all shots hit close to each other, but for some reason there is a systematic (consistent) error causing the shots to miss the centre; all measurements agree with each other, but they don't agree with the "true" value.

In case 3, it appears that our "average" shot has hit the bullseye, but no individual shot hits the bullseye. Repeated measurements may be of help to zero in on the "true" value.

In case 4, a larger lack of precision has prevented even our "average" from being useful.

Significant Digits

Rules

> - All non-zero digits are significant.
> - "Captive" zeros (zeros between non-zero digits) are significant.
> - Leading zeros (zeros on the left) are never significant.
> - Trailing zeros...
>   - a) are not significant if they are place holders (left of a decimal).
>   - b) are significant if they indicate a measurement (right of a decimal).
> - Exact numbers (counting numbers) have unlimited significant digits.

Examples

a) 200   one s.f. $= \underline{2} \times 10^2$
b) 2<u>02</u>   three s.f. $= \underline{2.02} \times 10^2$
c) 2<u>00.0</u>   four s.f. $= \underline{2.000} \times 10^2$
d) 0.0<u>2</u>   one s.f $= \underline{2} \times 10^{-2}$
e) 0.0<u>202</u>   three s.f. $= \underline{2.02} \times 10^{-2}$
f) 0.0<u>200</u>   three s.f. $= \underline{2.00} \times 10^{-2}$

Determine the number of significant figures in the following measurements...

a) 4010
b) 3.520
c) 0.00214
d) 0.01010
e) 10.02
f) 2.50×10^3

11.1 Learning Check

Rounding off Numbers

Rules for Rounding Off

> If the digit to be removed
> - a) is less than 5 the preceding digit stays the same
> - b) is 5 or greater, the preceding number is increased by 1.

DO NOT ROUND OFF RAW DATA; ONLY ROUND THE FINAL ANSWER.

Be Careful

Round the following to the indicated number of significant digits.

a) 2150 (2 s.d.)
b) 0.0346 (2 s.d.)
c) 1.994 (3 s.d.)
d) 0.0256 (2 s.d.)
e) 0.0050129 (3 s.d.)
f) 2149.99 (3 s.d.)

11.2 Learning Check

Adding and Subtracting Significant Digits

> Add up the digits by columns.
> Round off the final answer to the same number of **decimal places** as the least precise number.

Examples

Example 1	Example 2	Example 3	Example 4
41.11	0.2129	101.4	1.0
20.5	0.002	25	2.04
+18.333	+ 0.03	+201	+5.03
79.943	=0.2449	327.4	=8.07
=79.9	=0.24	=327	=8.1

Multiplying and Dividing Significant Digits

> Multiply the values as normal.
> Round off the final answer to the same number of **significant digits** as the least precise number.

Examples

Example 1	Example 2	Example 3
2.5 (2 sd)	8.314 (4 sd)	35.45 (4 sd)
x 1.1111 (5 sd)	x 2.5×10^{-2} (2sd)	x2.25 (3 sd)
=2.77775	=0.20785	=79.7625
=2.8 (2 sd)	= 0.21 (2sd)	=79.8 (3 sd)

Exam Hint

The easiest way not lose points on the exam is to write the answer with the same number of significant digits as the information in the question.

11.3 Learning Check

Carry out the following calculations and express your answer with the appropriate number of significant figures.

a) 2.50 + 3.369
b) 23.29 - 8.314
c) 0.0250 + 2.5×10^{-3}
d) 0.10 - 0.061
e) 8.314 x 101.325
f) 35.45 x 0.075
g) 2.50 ÷ 0.92
h) 101000 ÷ 8.3

Systematic Uncertainty

If a process or device always pushes an answer in a single direction, then the uncertainty is said to be systematic.

Consider a balance that is mis-calibrated, and always measures one gram too many; your results will be pushed in one direction. Reading a burette from above eye-level will cause the recorded data to be consistently to large.

Absolute & Relative Uncertainty

An uncertainty exists with every measurement.

The absolute uncertainty for any measuring device is the same for each measurement. This is why you can state it in the heading of a data table. On analogue devices (thermometers, glassware etc.) it is commonly taken as half of the smallest division.

Relative uncertainty is a measure of the uncertainty with respect to the magnitude of the measurement. It is usually expressed as a percentage with one or two significant digits.

Calculate the relative uncertainty by dividing the absolute uncertainty by the measurement and multiply by 100%. (You have to do it for each measurement; it's best to use a spreadsheet program)

Ideally we would like very low relative uncertainty. Then any uncertainty is insignificant compared to the measurement.

Propagating Uncertainty

The uncertainty of a calculation is always greater than the individual uncertainties.

> When **adding or subtracting** measurements, we add the **absolute** uncertainty.
> When **multiplying or dividing** measurements we add the **relative** uncertainty.

Let's take a practical example.

Example

> Determine the amount of heat required to raise the temperature of 75.00 g of water from 25.0°C to 85.5°C. The specific heat capacity of water is 4.18 J g^{-1}°C^{-1}. The uncertainties are ±0.005 g and ±0.5 °C.

Solution
mass = 75.00 g ± 0.005 g
T_1 = 25.1°C ± 0.5°C T_2 = 85.5°C ± 0.5°C

First of all we calculate ΔT = 60.4°C ± 1.0°C (add the abs. uncert.)

The next step requires us to use relative uncertainties. We convert the absolute uncertainties into relative percent uncertainty.

mass: 0.005 ÷ 75.00 × 100% = 0.007% : 75.00 ± 0.007%
ΔT: 0.4 ÷ 60.4 × 100% = 1.65% = 2% : 60.4°C ± 2%

$Q = mc\Delta T$
Q = 75.00g ± 0.007% × 4.18 J g^{-1}°C^{-1} × 60.4°C ± 2%
Q = 18935.4 J ± 2.007%
Q = 18900 J ± 2%   We only report 1 sd in the uncertainty.
Q = 18900 J ± 400 J   We may want to convert back to absolute
Q = 18.9 kJ ± 0.4 kJ

In this example the uncertainty of ±0.5°C on the thermometer is much greater than the uncertainty on the balance. You can ignore the uncertainty of the balance.

Types of Relationships

Graphs tend to be linear or adjusted through mathematical manipulations so that they are in the form of a straight line. This allows us to determine the intercepts and the slope which may have a significance.

Examples

The "Charles' Law" graph uses the x-intercept to find absolute zero.

The Arrhenius plot manipulates the equation so that we can use the slope of the line to determine the activation energy of a reaction.

The Rate of reaction graphs can be used to calculate the rate at an instant by finding the slope of the tangent of the curve at an instant.

Another way to determine the validity of results is through correlation.

Correlation is how close each point follows a trend line, and is a measure of the accuracy and precision of your data.

Case 1	Case 2	Case 3	Case 4
High Correlation Positive slope	High Correlation Negative slope	Low Correlation	No Correlation

In cases 1 & 2, the points all fall on the line, correlation is high.

In case 3, the points fall near the line, but not on it - correlation is low, but it does exist.

In case 4, which line is correct? - There is no clear trend. There is no correlation.

The Formula of a compound

We can combine two pieces of information already learned to help determine the formula and the properties of a compound.

The empirical and molecular formulae can tell us the relative and actual ratio of atoms in a compound. The general formula from organic chemistry can tell us about the degree of saturation.

For organic molecules, we know that they are saturated with hydrogen if the number of hydrogens is double plus two compared to the carbon atoms. Each pair of hydrogens that are missing represent either a pi bond or a ring closure. The Index of Hydrogen Deficiency (IHD) gives us a clue.

Formula	Structure		comment / IHD
C_6H_{14}	hexane		saturated C_nH_{2n+2}
C_6H_{12}	hex-2-ene		1 double bond C_nH_{2n}
	cyclohexane		ring C_nH_{2n}
C_6H_{10}	hex-2,4-diene		C_nH_{2n-2} 2 double bonds
	hex-2-yne		C_nH_{2n-2} 1 triple bond
	cyclohexene		C_nH_{2n-2} ring and a double bond
C_6H_8	hexatriene		
	cyclohexadiene		
C_6H_6	1,3,5-cyclohexatriene benzene		

Formula	Structure		comment / IHD
$C_5H_{12}O$	pentanol	$H_3C-CH_2-CH_2-CH_2-CH_2-OH$	ratio of C:H is saturated $C_nH_{2n+2}O$
$C_5H_{10}O$	pentanal / pentanone pentenol	$H_3C-CH_2-CH_2-CH_2-CHO$ $H_3C-CH_2-C(=O)-CH_2-CH_3$ $H_2C=CH-CH_2-CH_2-CH_2-OH$	typically the C=O will be the double bond, unless the C=C and the OH are far apart in the molecular structure - the latter is an "enol" (ene C=C; ol OH)
C_5H_8O	cyclopentanone	(cyclopentanone ring structure)	

Mass Spectroscopy (MS)

Mass spectroscopy is a technique of measuring the masses of atoms, molecules or molecular fragments.

When an atom or molecule enters the MS it is bombarded with electrons, causing a molecular electron to be knocked off and a positive ion formed. Some of these positive ions are more stable than others, and unstable molecular ions can break into fragments. The fragments will either be positive ions or neutral radicals. Only the positive ions move through the devices electromagnetic fields in a measurable way. Knowing this, we can work out the pieces of the molecule and have an idea of what structure it has.

Nuclear Magnetic Resonance

In nuclear magnetic resonance (NMR) the nuclei of atoms are analysed, and of interest to us is the arrangement of protons (hydrogen). We will look at proton-NMR or ^1H NMR.

The idea is that the protons behave like little spinning tops. In a strong magnetic field the tops are all spinning in line with the external magnetic field. We can give them some energy (in the form of radio waves) to flip their spin states to be antiparallel to the external field. We do this to all protons, and then "listen" to the radio waves emitted as the protons return to their lower spin state. The neat thing is that protons that are in the same chemical environment release the same amount of radio energy.

The NMR 1H spectrum gives us four different pieces of information

1. The "Chemical Shift" measured in ppm tells us how close to an electronegative atom (like O or Cl) a hydrogen environment is.

2. The number of peaks tells us the number of groups of equivalent hydrogens.

3. The integration (the area under the peak) proportional to the number of hydrogens in a particular environment. Be

careful, because sometimes there isn't a single hydrogen, so the actual ratio could be higher.

4. The splitting pattern tells us the number of neighbours. A peak is split into the number of neighbours plus one. So a neighbouring CH2 group, will cause a triplet.

The reference for the graph is tetramethylsilane, TMS; Si(CH3)4, which contains 12 identical hydrogens in a very non-polar molecule. This is the zero reference point.

Consider the molecule CH3CH3CHBrOCH3, the NMR spectrum may look something like this.

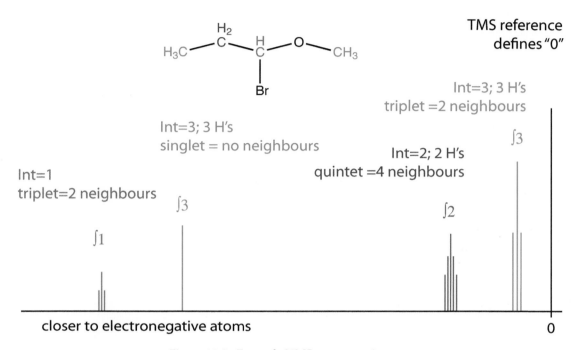

Figure 11.2: Example NMR proton assignment

Infrared Spectroscopy

We have often thought of a chemical bond as a type of mechanical spring holding atoms together. Infrared radiation can cause these springs to vibrate (stretch or bend) depending upon the strength of the bond. Each type of bond has its own strength and therefore vibrational frequency. Therefore different types of bonds will absorb different amounts of infrared energy, and we can measure that absorbance. However, causing one bond to vibrate will set off a chain reaction (like plucking a spiderweb) and other bonds will start to vibrate as the energy gets passed around the molecule. This is the basis of the "fingerprint" region - a set of low energy vibrations unique to the molecule.

X-ray Crystallography

X-ray crystallography was developed by a father and son (Bragg) and they are the only father-son team to win a Nobel Prize (so far).

The process uses x-rays and a phenomena known as "diffraction" to measure the bond lengths in a crystal. If you can obtain a crystal of a pure substance, you can aim x-rays at it, and use the resulting image / pattern to work out the bond lengths. This was one of the pieces of evidence that helped to discover the structure of DNA.

Answers

Learning Check 1.1

- a) 1.2×10^{24}
- b) 7.53×10^{22}
- c) 9.03×10^{23}
- d) 7.53×10^{20}
- e) 7.53×10^{18}
- f) 2.74×10^{15}
- g) 3.2×10^{24}
- h) 1.11×10^{22}

Learning Check 1.2

- a) 100 mol
- b) 0.375 mol
- c) 5.00 mol
- d) 1.25×10^{-3} mol
- e) 5.75×10^{-6} mol
- f) 8.75×10^{-5} mol
- g) 4.25×10^{-4} mol
- h) 1.74 mol

Learning Check 1.3

- a) 58.44 g mol^{-1}
- b) 155.99 g mol^{-1}
- c) 80.06 g mol^{-1}
- d) 110.98 g mol^{-1}
- e) 310.98 g mol^{-1}
- f) 246.51 g mol^{-1}
- g) 159.70 g mol^{-1}
- h) 149.12 g mol^{-1}
- i) 60.06 g mol^{-1}

Learning Check 1.4

- a) 63.6 g
- b) 167 g
- c) 22.2 g
- d) 7.65×10^{-4} g
- e) 0.415 g
- f) 4.03×10^{-6} g
- g) 90.3 g
- h) 6.37×10^{-4} g
- i) 245.2 g

Learning Check 1.5

- a) 6.27×10^{-2} mol
- b) 8.31×10^{-2} mol
- c) 7.04×10^{-2} mol
- d) 1.38×10^{-2} mol
- e) 3.00×10^{-2} mol
- f) 6.86×10^{-1} mol
- g) 9.885×10^{-2} mol
- h) 1.86×10^{-2} mol
- i) 4.97×10^{-3} mol

Learning Check 1.6

a) $2K + 2H_2O \rightarrow H_2 + 2KOH$

b) $3CuO + 2NH_3 \rightarrow 3H_2O + N_2 + 3Cu$

c) $2Al + 6HCl \rightarrow 2AlCl_3 + 3H_2$

d) $2ZnS + 3O_2 \rightarrow 2ZnO + 2SO_2$

e) $2NH_4Cl + Ca(OH)_2 \rightarrow CaCl_2 + 2NH_3 + 2H_2O$

f) $2C_4H_{10} + 13O_2 \rightarrow 8CO_2 + 10H_2O$

g) $H_3PO_4 + 3NaOH \rightarrow Na_3PO_4 + 3H_2O$

h) $2C_2H_2 + 5O_2 \rightarrow 4CO_2 + 2H_2O$

i) $2KClO_3 \rightarrow 2KCl + 3O_2$

j) $2C_2H_6 + 7O_2 \rightarrow 4CO_2 + 6H_2O$

k) $N_2 + 3H_2 \rightarrow 2NH_3$

l) $N_2H_4 + O_2 \rightarrow N_2 + 2H_2O$

m) $2Na + Cl_2 \rightarrow 2NaCl$

n) $4Fe + 3O_2 \rightarrow 2Fe_2O_3$

o) $2SO_2 + O_2 \rightarrow 2SO_3$

Learning Check 1.7

1. a) 2.5 mol   b) 7.5 mol
2. a) i) 15 mol   ii) 100 mol
       iii) 2.5 mol
   b) 10.0 mol
   c) i) 12 mol   ii) 80 mol
       iii) 2.0 mol
3. a) 62.5 mol
   b) 0.80 mol
   c) i) 2.48 mol   ii) 0.275 mol

Learning Check 1.8

1. a) 2500 g of oxygen
   b) 558 g of ethanol
2. 703 g of iron(III) oxide
3. 1.72×10^{-2} g of silver sulphide
4. a) 2.86 kg of oxygen
   b) 3.29 kg of NO_2
5. a) 198 g of tungsten
   b) 6.53 g of hydrogen required.

Learning Check 1.9

1. a) n = 0.339 mol of ethane
       n = 1.39 mol of oxygen
   b) C_2H_6 is the limiting reactant
   c) 29.8 g of CO_2 produced

2. n=0.9 mol of NaOH
   n=0.122 mol of H_3PO_4
   a) H_3PO_4 is the limiting reactant
   b) 20.1 g of Na_3PO_4 are produced
3. a) $2C_4H_{10} + 13O_2 \rightarrow 8CO_2 + 10H_2O$
   b) n=3.125 mol of O_2
      n=1.72 mol of C_4H_{10}
   c) O_2 is the limiting reactant
   d) 84.63 g of CO_2 formed
      43.32 g of H_2O formed

Learning Check 1.10
a) 1.50 M
b) 1.71 M
c) 0.0200 M
d) 0.100 M
e) 41.5 g
f) 4.25 g
g) 1.00 g
h) 29.7 g

Learning Check 1.11
1. 88%
2. 81.77%

Learning Check 1.12
1. C_2H_5
2) C_3H_4
3. CH

Learning Check 1.13
1. CH_3
2. CH_2
3. C_3H_8

Learning Check 1.14
1. CH_2O
2. C_3H_8O
3. C_2H_4O

Summary Questions - Chapter 1
1. a) $AgNO_3(aq) + NaCl(aq) \rightarrow$
   $AgCl(s) + NaNO_3(aq)$
   b) 3.00×10^{-3} mol AgCl(s)
   c) 0.15 mol dm^{-3}
   d) 2.19 g

2. a) CH
   b) 1.544×10^{-3} mol
      26.22 g mol^{-1}
   c) C_2H_2
   d) $CaC_2(s) + 2H_2O(l) \rightarrow$
      $Ca(OH)_2(aq) + C_2H_2(g)$

3. a) $n(C_7H_6O_3) = 4.34 \times 10^{-2}$ mol
      $n(C_4H_6O_3) = 4.90 \times 10^{-2}$ mol
   b) $C_7H_6O_3$ is the limiting reactant.
   c) 7.83 g of aspirin
   d) 83.5%

4. a) $2C_6H_{14} + 19O_2 \rightarrow 12CO_2 + 14H_2O$
   b) 5.22×10^{-2} mol hexane
      3.516×10^{-1} mol oxygen
   c) O_2 is the limiting reactant
   d) 9.77 g of CO_2 produced
      4.67 g of H_2O produced
   e) 1.41 g of unreacted C_6H_{14}

5. a) 5.000×10^{-3} mol of $BaSO_4(s)$
   b) 0.2000 mol dm^{-3}
   c) 0.1000 mol in 0.500 dm^3
   d) 109.9 g mol^{-1}
   e) Li

Answers

CHAPTER 2
Learning Check 2.1

	A	Z	p	n	e
$^{1}_{1}H$	1	1	1	0	1
$^{2}_{1}H$	2	1	1	1	1
$^{3}_{1}H$	3	1	1	2	1
$^{10}_{5}B$	10	5	5	5	5
$^{11}_{5}B$	11	5	5	6	5
$^{35}_{17}Cl$	35	17	17	18	17
$^{37}_{17}Cl$	37	17	17	20	17

Learning Check 2.2
1. 107.96 = 108.0
2. 24.3
3. 10.80

Learning Check 2.3
1. 31.5% ^{203}Tl; 68.5% ^{205}Tl
2. 6.00% ^{6}Li and 94.0% ^{7}Li
3. 64% ^{69}Ga; 36% ^{71}Ga

Learning Check 2.4
N $1s^2 2s^2 2p^3$

S $1s^2 2s^2 2p^6 3s^2 3p^4$

Mg $1s^2 2s^2 2p^6 3s^2$

Al^{3+} $1s^2 2s^2 2p^6$

Cl^- $1s^2 2s^2 2p^6 3s^2 3p^6$

Cu^+ $1s^2 2s^2 2p^6 3s^2 3p^6 4s^1 3d^{10}$

Learning Check 2.5
1. Red = 2.84×10^{-19} J

   Blue = 4.97×10^{-19} J
2. 549.4 nm

Chapter 2 Summary Questions
1. 109.5
2. 35% ^{89}Lm, 65% ^{90}Lm
3. a) $1s^2 2s^2 2p^6 3s^2 3p^3$
   b) $1s^2 2s^2 2p^6 3s^2 3p^6 4s^2 3d^{10} 4p^4$
   c) $1s^2 2s^2 2p^6 3s^2 3p^6 4s^2 3d^{10} 4p^6$
   d) $1s^2 2s^2 2p^6 3s^2 3p^6$
   e) $1s^2 2s^2 2p^6 3s^2 3p^6 4s^0 3d^8$
4. a)

↑↓		↑	↑	
$2s^2$			$2p^2$	

b)

↑↓		↑↓	↑	↑
$3s^2$			$3p^4$	

c)

↑↓		↑	↑	↑		
$4s^2$			$3d^3$			

Learning Check 3.1

Larger	Explanation
O^{2-}	Increased e^-/e^- repulsion or lower net attractive forces for same # of electrons
Mg	3 shells of electrons vs. 2 shells of electrons
F^-	Increased e^-/e^- repulsion or lower net attractive forces for same # of electrons
K	4 shells of electrons vs. 3 shells of electrons
Ar	Lower nuclear charge on Ar, for same # of electrons
Na	3 shells in Na vs. 2 shells in Li
N	N has lower nuclear charge, therefore lower attraction between nucleus & e^-

Learning Check 3.2
a) Be, Mg, Ca   b) Xe, I, Te

c) Ge, Ga, In   d) F, N, As

e) F, Cl, S   f) Li, K, Cs

Learning Check 3.3

a) Li – fewest shells

b) P – fewer shells

c) O$^+$ - least e-/e- repulsion

d) Cl – greatest #protons for fewest shells

e) Ni – fewest shells

f) Ar – all have same # electrons, Ar has most protons

Learning Check 3.4

a) Mg$^+$ - still has 3s^1 electron, Mg^{2+} is 2p^{6}3s^0

b) O$^-$ - more e-/e- repulsion

c) Ar, more e-/e- repulsion

d) Mg$^+$ - still has 3s^1 electron, Na$^+$ is 2p^6

e) Al^{2+} - still has 3s^1 electron, Na$^+$ is 2p^6

Learning Check 3.5

a) Ca, Mg, Be   b) Te, I, Xe

c) In, Ga, Ge   d) As, N, F

e) S, Cl, F   f) Cs K, Li

Chapter 3 Summary Questions

1 a) Sulphur is 3p^4, and the fourth electron is easier to remove due to electronic repulsion in the p orbital with two electrons.

b) The electron in aluminium is further away from the nucleus

c) You are trying to remove a 2p^6 electron from Na$^+$ and only a 3s^1 electron from Mg$^+$

2. a) The I.E. decreases down the group. The greater number of shells means that the attractive force between the nucleus and the valence electron is less, so less energy is required to remove it

b) The I.E. increases across a period due to an increasing nuclear charge, and decreasing atomic radius

4. a) 2Li + I$_2$ → 2LiI

b) 2K + 2H$_2$O → 2KOH + H$_2$

c) Br$_2$ + 2I$^-$ → 2Br$^-$ + I$_2$

5. Ti^{2+} = 1s^{2}2s^{2}2p^{6}3s^{2}3p^{6}4s^{0}3d^2

Ti^{4+} = 1s^{2}2s^{2}2p^{6}3s^{2}3p^6

Ti^{2+} should be colourful due to presence of the two d-electrons

6. Cu^{2+} ions are colourful due to the partially filled (d^9) d-shell.

Zn^{2+} has a full d shell.

Learning Check 4.1

a) Na$_2$S   b) BeF$_2$

c) GaI$_3$ d) K$_3$N

e) AlP   f) Mg$_3$N$_2$

Learning Check 4.2

a) copper(I) chloride

b) cobalt(II) iodide

c) chromium(III) oxide

d) nickel(III) bromide

e) manganese(IV) oxide

f) copper(1) sulphide

Learning Check 4.3

a) FeO   b) HgCl$_2$

c) Cu$_2$O   d) Ni$_2$O$_3$

e) CoCl$_2$ f) Fe$_2$S$_3$

Learning Check 4.4

a) NH$_4$Cl   b) NaNO$_3$

c) K$_2$CO$_3$ d) CaSO$_4$

e) Mg$_3$(PO$_4$)$_2$ f) (NH$_4$)$_2$CO$_3$

Learning Check 4.5

a) sodium hydrogen carbonate

b) sodium nitrite

c) ammonium nitrate

d) lithium phosphate

e) barium sulphate

f) ammonium sulphate

Answers

Learning Check 4.6

a) FeSO$_4$
b) Ni(NO$_3$)$_3$
c) Cr(NO$_3$)$_6$
d) Cu$_2$CO$_3$
e) Hg(NO$_3$)$_2$
f) MnSO$_4$

Learning Check 4.7

a) nickel(II) carbonate
b) copper(II) sulphate
c) iron(III) sulphate
d) cobalt(II) nitrate
e) nickel(III) nitrate
f) mercury(I) carbonate

Learning Check 4.8 & 4.9

a) trigonal pyramid

b) tetrahedral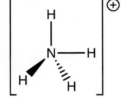

c) tetrahedral

d) trigonal planar

e) bent / V-Shape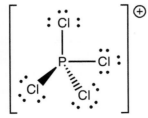

f) tetrahedral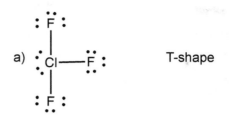

Learning Check 4.10 & 4.11

a) T-shape

b) 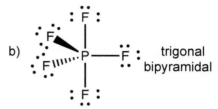 trigonal bipyramidal

c) 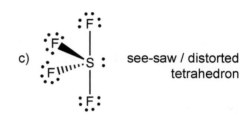 see-saw / distorted tetrahedron

d) 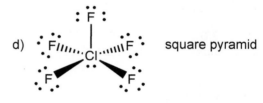 square pyramid

e) 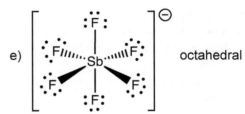 octahedral

f) 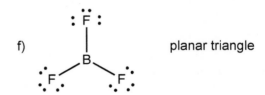 planar triangle

Learning Check 4.12

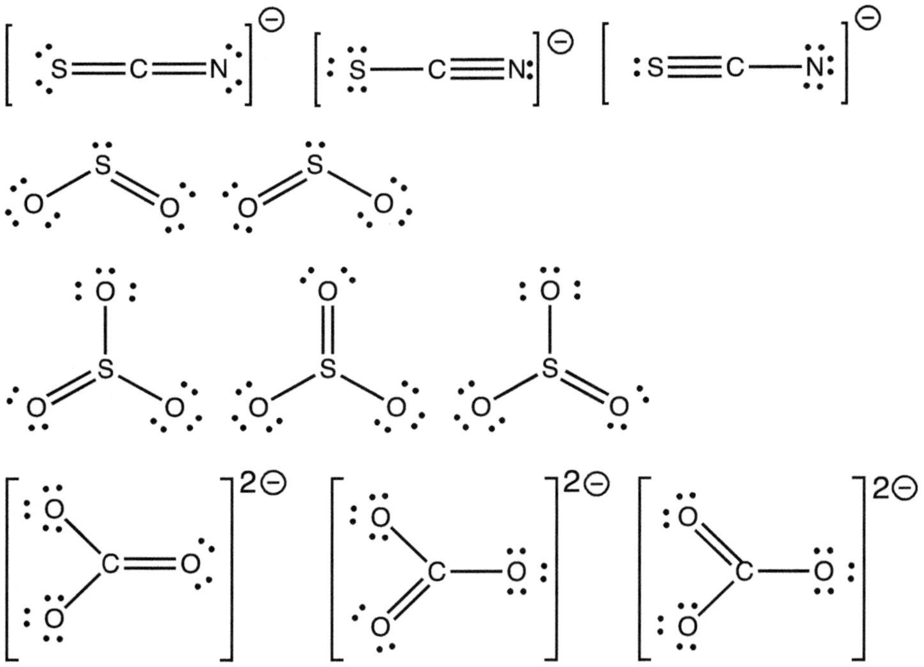

[Lewis structures for SCN⁻ resonance forms, SO₂ resonance forms, SO₃ resonance forms, and CO₃²⁻ resonance forms]

Learning Check 4.13

Bond orders are:

SCN⁻ 2.0   SO$_2$ 1.5

SO$_3$ 1.3   CO$_3^{2-}$ 1.3

Learning Check 4.14

molecule	hybridization	shape	angle
CH$_3$CH$_3$	sp^3	tetrahedral	109.5°
CH$_2$CH$_2$	sp^2	trigonal planar	120°
O$_2$	sp^2	linear	not relevant - 2 atoms
CH$_3$CHO	sp^2	trigonal planar	120°
CH$_3$COCH$_3$	sp^2	trigonal planar	120°
CH$_3$CN	sp	linear	180°
N$_2$	sp	linear	not relevant - 2 atoms
CH$_3$NH$_2$	sp^3	tetrahedral	107° (one lone pair)

Learning Check 4.15

Molecular Structure	Hybridization	Bond Angle	Type of Bond	Shape
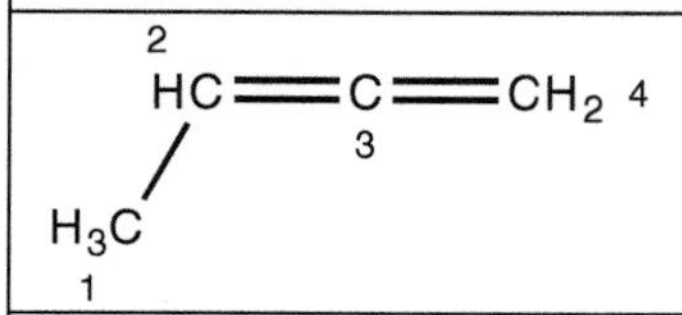 HC=C=CH$_2$ with H$_3$C (positions 1=H₃C, 2=HC, 3=C, 4=CH₂)	1 sp^3 2 sp^2 3 sp 4 sp^2	109.5° 120 180 120	single double 2 double double	tetrahedral trigonal planar linear trigonal planar
H$_3$C—C≡C—CH$_3$ (positions 1, 2, 3, 4)	1 sp^3 2 sp 3 sp 4 sp^3	109.5 180 180 109.5	single triple triple single	tetrahedral linear linear tetrahedral

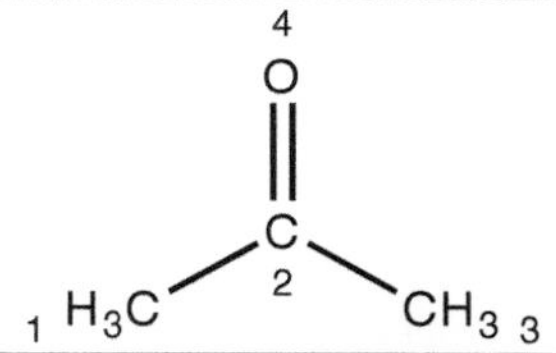

Answers

Structure	Atom	Hybridization	Angle	Bond	Geometry
$_1H_3C-\overset{4}{\underset{2}{C}}(=O)-CH_{3\ 3}$	1 sp³	109.5	single	tetrahedral
	2 sp²	120	double	trigonal planar
	3 sp²	120	double	trigonal planar
	4 sp³	109.5	single	tetrahedral
$_1H_3C-\underset{2}{C}(=\overset{4}{O})-\underset{3}{OH}$	1 sp³	109.5	single	tetrahedral
	2 sp²	120	double	trigonal planar
	3 sp²	120	double	trigonal planar
	4 sp³	109.5	single	tetrahedral

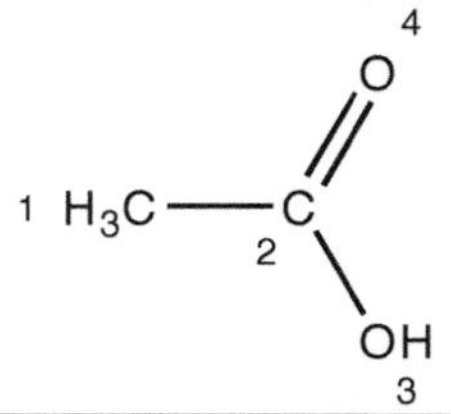

Chapter 4 Summary

1. a) calcium nitrate
    b) copper(II) nitrate
    b) aluminium sulphate
    d) iron(II) phosphate
    e) ammonium chloride
    f) sodium hydrogen carbonate

2. a) K₂O   b) NH₄NO₃
    c) MgCO₃   d) Ni₂O₃
    e) PbO₂   f) AlPO₄

3. a) :C≡O:

    b) N with two I and one lone pair (pyramidal)

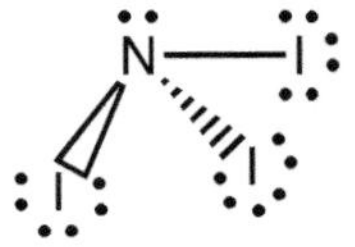

    c) CH₂Cl₂ tetrahedral

    d) [O-N-O]⁻ nitrite ion

    e) [O=N=O]⁺

    f) :F̈—Kr—F̈:

    g) IF₃ (T-shaped) with F axial and equatorial

    h) SeI₄ (seesaw)

    i) AsF₅ (trigonal bipyramidal)

    j) SF₆ (octahedral)

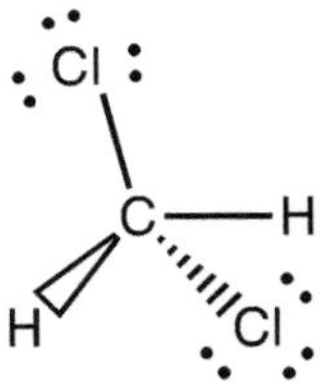

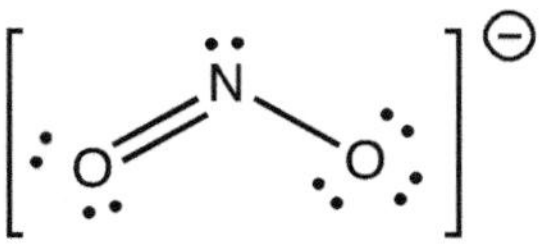

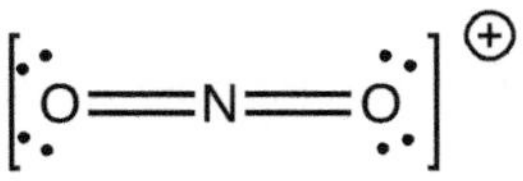

page 197

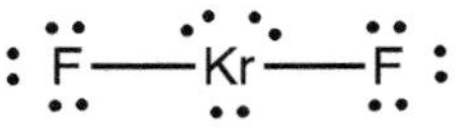

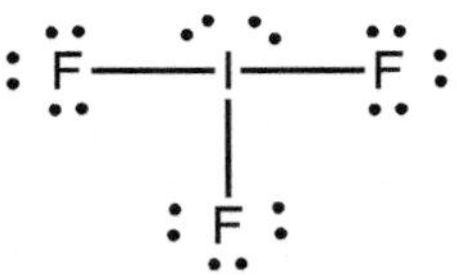

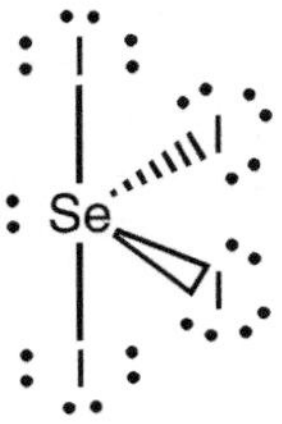

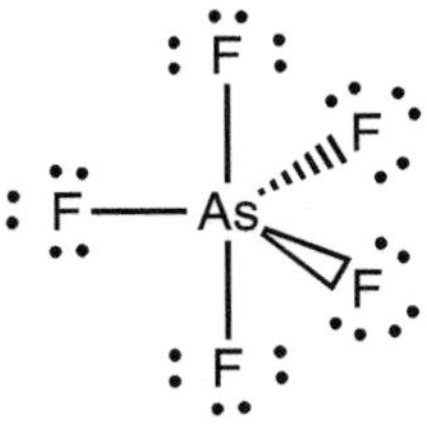

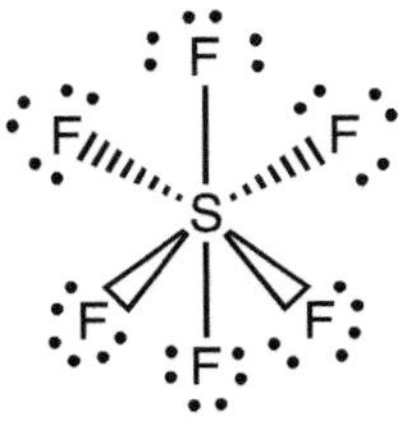

4.
   a) Hydrogen bonding
   b) Dipole forces
   c) Dipole forces
   d) Dipole forces
   e) Hydrogen bonging
   f) Dipole forces

5. $[\ddot{\underset{..}{O}}=C=\ddot{N}\!:\,]^{\ominus}$

   $[\,:O\equiv C-\ddot{\underset{..}{N}}:\,]^{\ominus}$

   $[\,:\ddot{\underset{..}{O}}-C\equiv N:\,]^{\ominus}$

6. NCl_3 has the higher melting point because of the dipole forces resulting from the asymmetry in the trigonal pyramidal geometry, due to its lone pair. BCl_3 is a planar triangle nonpolar molecule and therefore has a lower boiling point due to its Van der Waal's forces.

7. Propene contains three carbons atoms. One carbon atom -CH_3, is sp^3 hybridized, and has a tetrahedral geometry and has four sigma bonds. The other two carbons are sp^2 hybridized with planar triangular geometry and have 3 sigma bonds. There is also a pi bond connected these two carbons.

Learning Check 5.1

1. -134.8 kJ
2. -171.07 kJ
3. -98 kJ
4. -296.1 kJ
5. 226 kJ

Learning Check 5.2

1. -808 kJ mol^{-1}
2. -109 kJ mol^{-1}
3. -577 kJ mol^{-1}
4. -184 kJ mol^{-1}
5. -859 kJ mol^{-1}

Learning Check 5.3

1. -198 kJ mol^{-1}
2. -179.7 kJ mol^{-1}
3. -99.6 kJ mol^{-1}
4. +155.0 kJ mol^{-1}
5. -853.8 kJ mol^{-1}

Learning Check 5.4

a. -206 kJ mol^{-1}
b. +230 kJ mol^{-1}

Learning Check 5.5

1. positive (forming a gas)
2. positive (mixing)
3. negative (solid formed)
4. positive (3 more moles of gas)
5. zero / negilible (equal moles of gas)
6. negative (1 mole of gas less)

Chapter 5 Summary Questions

1. +322.2 kJ
2. -703 kJ
3. -14.2 kJmol^{-1}
4.
   a) no change
   b) double ($\Delta T = 6.8°C$)
   c) no change
   d) less ($\Delta T = 2.27°C$)
5.
   a) Positive (endothermic)
   b) Positive (forming a gas)
   c) increase spontaneity
   d) 60°C

Chapter 6 Summary Questions

1.
   a) First order
   b) First order
   c) Rate = $k[S_2O_8^{2-}][I^-]$
   d) k = 0.412 mol^{-1}·dm3·s^{-1}

Answers

2. a) i) Rate = $k[H_2][ICl]$

   ii) Rate = $k[H_2][ICl]^2$

   iii) Rate = $k[ICl]^2$

   b) The elementary step would involve three species colliding which is very unlikely.

Learning Check 7.1

1. $K_c = \dfrac{[H_2O]^2[SO_2]^2}{[H_2S]^2[O_2]^3}$

2. $K_c = \dfrac{[NO_2]^4[O_2]}{[N_2O_5]^2}$

3. $K_c = \dfrac{[CH_3OH]}{[CO][H_2]^2}$

4. $K_c = \dfrac{[N_2]^2[H_2O]^6}{[NH_3]^4[O_2]^3}$

5. $K_c = \dfrac{[NO_2]^4}{[N_2O]^2[O_2]^3}$

Learning Check 7.2

Learning Check 7.3

1. $K_c = 0.052$

2. $K_c = 0.398$

3. $[N_2O_5] = [NO] = 7.1 \times 10^{-7}$

Summary Questions Chapter 7

1. a) $K_c = \dfrac{[NO_2]^2}{[N_2O_4]}$

   b) $K_c = \dfrac{[SiCl_4][H_2]^2}{[SiH_4][Cl_2]^2}$

   c) $K_c = \dfrac{[PCl_3]^2[Br_2]^3}{[PBr_4]^2[Cl_2]^3}$

   d) $K_c = \dfrac{[CH_3OH]}{[CO][H_2]^2}$

   e) $K_c = \dfrac{[NO]^2[O_2]}{[NO_2]^2}$

2. a) Colour darkens as equilibrium shifts right to reduce the number of moles of gas.

   b) The reverse reaction is endothermic, therefore the colour lightens.

   c) A catalyst has no effect as it lowers the activation energy of the forward and reverse reactions equally.

   d) Addition of a noble gas has no effect.

   e) The equilibrium will shift right to use up the excess oxygen added.

3. a) $[N_2O] = 0.010$ M

   $[O_2] = 0.041$ M

   b) 23.2

4. $[H_2] = 5.3 \times 10^{-7}$

   $[O_2] = 2.6 \times 10^{-7}$

Learning Check 8.1

a) $HNO_3(aq) + NaHCO_3(aq) \rightarrow NaNO_3(aq) + H_2O(l) + CO_2(g)$

b) $Al_2O_3(s) + 6HCl(aq) \rightarrow 2AlCl_3(aq) + 3H_2O(l)$

c) $ZnO(s) + H_2SO_4(aq) \rightarrow ZnSO_4(aq) + H_2O(l)$

d) $Mg(s) + 2HNO_3(aq) \rightarrow Mg(NO_3)_2(aq) + H_2(g)$

e) $H_2SO_4(aq) + CuCO_3(s) \rightarrow CuSO_4(aq) + H_2O(l) + CO_2(g)$

f) $2HCl(aq) + Ca(OH)_2(aq) \rightarrow CaCl_2(aq) + 2H_2O(l)$

Learning Check 8.2

1. a) HF   b) $N_2H_5^+$

   c) $C_5H_6N^+$ d) HO_2^-

   e) H_2CrO_4 f) H_2O_2

2. a) NH_2^- b) CO_3^{2-}

   c) CN^- d) $H_4IO_6^-$

   e) NO_3^- f) OH^-

3. a) acid, base $\rightleftharpoons$ conjugate base, conjugate acid

   b) base, acid $\rightleftharpoons$ conjugate acid, conjugate base

   c) acid, base $\rightleftharpoons$ conjugate base, conjugate acid

   d) acid, base $\rightleftharpoons$ conjugate base, conjugate acid

Learning Check 8.3

1. a) pH = 0.70 b) pH = 1.92
2. a) 1.30 b) 5.25
3. a) 2.40 b) 4.93

Learning Check 8.4

pH = 9.43

Chapter 8 Summary

1. a) $2HNO_3(aq) + CuO(s) \rightarrow Cu(NO_3)_2(aq) + H_2O(l)$

   b) $2Al(s) + 6HCl(aq) \rightarrow 2AlCl_3(aq) + 3H_2(g)$

   c) $Fe_2(CO_3)_3(s) + 3H_2SO_4(aq) \rightarrow Fe_2(SO_4)_3(aq) + 3H_2O(l) + 3CO_2(g)$

2. a) HSO_4^- b) HCO_3^-
   c) $C_2H_4O_2^-$ d) NH_3
   e) NH_4^+ f) $H_2PO_4^-$

3. a) I^- b) H^-
   c) NH_3 d) NO_2^-
   e) HPO_4^{2-} f) $H_2PO_4^-$

4. a) acid, base ⇌ conj. acid, conj. base
   b) base, acid ⇌ conj. acid, conj. base
   c) base, acid ⇌ conj. acid, conj. base
   d) acid, base ⇌ conj. base, conj. acid

5. a) pH = 1
   b) pH = 3
   c) pH = 7
   d) pH = 13
   e) pH = 7

Learning Check 9.1

+4	+3	+3	+3
MoS_2	Ni_2O_3	P_4O_6	As_2O_3

+3	+3	+2	+6
$Cr(NO_3)_3$	$Cr_2(SO_4)_3$	$CrSO_4$	$Cr(SO_4)_3$

+1	+3	+5	+7
ClO^-	ClO_2^-	ClO_3^-	ClO_4^-

Learning Check 9.2

1. Yes
2. Yes
3. Yes
4. No
5. No

Learning Check 9.3

1. H_2 is the reducing agent, Cl_2 is the oxidizing agent

2. MnO_2 is the oxidizing agent, the chloride ion is the reducing agent

3. CH_4 is the reducing agent, O_2 is the oxidizing agent.

Learning Check 9.4

a) 0.63 V
b) 1.25 V
c) 1.10 V

Learning Check 9.5

a) not spontaneous
b) not spontaneous
c) not spontaneous
d) spontaneous
e) MnO_4^-

Learning Check 9.6

Electrolyte	Cathode Product	Anode Product
NaCl(l)	Na(l)	$Cl_2(g)$
NaCl(aq) dilute	$H_2(g)$	$O_2(g)$
NaCl(aq) conc.	$H_2(g)$	$Cl_2(g)$
$CuSO_4$(aq)	Cu(s)	$O_2(g)$
Na_2SO_4(aq)	$H_2(g)$	$O_2(g)$
$MgBr_2$(aq)	$H_2(g)$	$Br_2(g)$
$PbBr_2$(l)	Pb(s)	$Br_2(g)$
$PbBr_2$(aq)	Pb(s)	$Br_2(g)$

Learning Check 9.7

Answers

a) $Zn(s) + 2HCl(aq) \rightarrow Zn^{2+}(aq) + 2Cl^-(aq) + H_2(g)$

b) $3I^-(aq) + ClO^-(aq) + 2H^+(aq) \rightarrow I_3^-(aq) + Cl^-(aq) + H_2O(l)$

c) $3As_2O_3(s) + 4NO_3^-(aq) + 4H^+(aq) + 7H_2O \rightarrow 6H_3AsO_4(aq) + 4NO(g)$

d) $10Br^-(aq) + 2MnO_4^-(aq) + 16H^+(aq) \rightarrow 5Br_2(aq) + 2Mn^{2+}(aq) + 8H_2O(l)$

e) $3CH_3OH(aq) + Cr_2O_7^{2-}(aq) + 8H^+(aq) \rightarrow 3CH_2O(aq) + 2Cr^{3+}(aq) + 7H_2O(l)$

Summary Questions Chapter 9

1.

+2	+4	+1	+2
CO	CO_2	Hg_2Cl_2	HgO

+7	+5	+6	0
$KMnO_4$	$Mg_2P_2O_7$	$XeOF_4$	As_4

+6	+2	+3	0
$Na_2C_2O_4$	$Na_2S_2O_3$	$HAsO_2$	S_8

2.
a) $2Fe^{2+}(aq) + 2H^+(aq) + H_2O_2(aq) \rightarrow 2Fe^{3+}(aq) + 2H_2O(l)$
b) $2S_2O_3^{2-}(aq) + I_2(aq) \rightarrow S_4O_6^{2-}(aq) + 2I^-(aq)$

3.

Learning Check 10.1

a) 2,4-dimethylhexane
b) 3,4-dimethylhexane
c) 2-methylpentane
d) 2,2-dimethylbutane
e) 2,3-dibromopentane
f) 2,2-dibromopentane
h) 1,2-dichlorobutane

Learning Check 10.2

a) Three isomers of C_5H_{12}

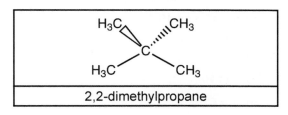

n-pentane

2-methylbutane

2,2-dimethylpropane

b) Five isomers of C_6H_{14}

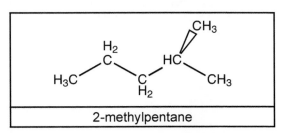

n-hexane

2-methylpentane

3-methylpentane

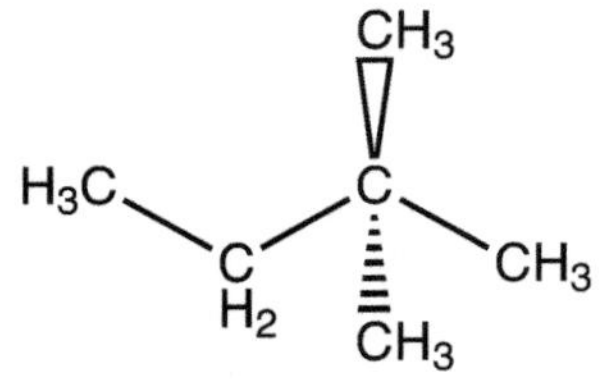

2,2-dimethylbutane

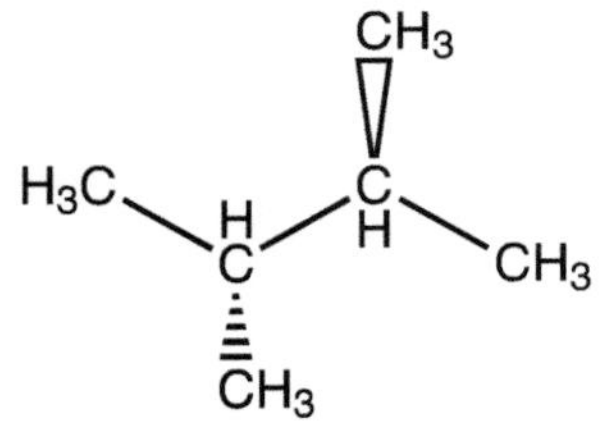

2,3-dimethylbutane

Learning Check 10.3

$CH_4 + 2O_2 \rightarrow CO_2 + 2H_2O$
$2C_2H_6 + 7O_2 \rightarrow 4CO_2 + 6H_2O$
$C_3H_8 + 5O_2 \rightarrow 3CO_2 + 4H_2O$
$2C_4H_{10} + 13O_2 \rightarrow 8CO_2 + 10H_2O$

Learning Check 10.4

$2CH_4 + 3O_2 \rightarrow 2CO + 4H_2O$
$2C_2H_6 + 5O_2 \rightarrow 4CO + 6H_2O$
$2C_3H_8 + 7O_2 \rightarrow 6CO + 8H_2O$
$2C_4H_{10} + 9O_2 \rightarrow 8CO + 10H_2O$

Learning Check 10.5

isomers of $C_4H_8O_2$

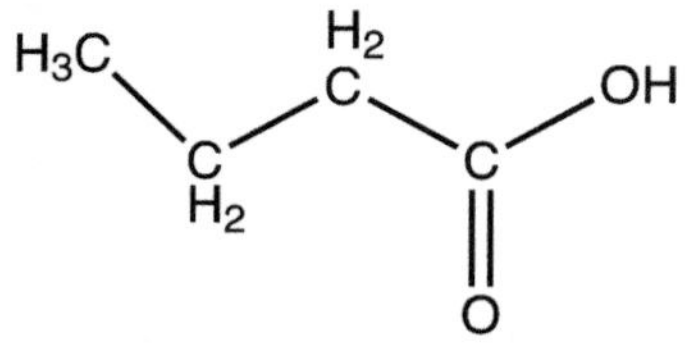

butanoic acid

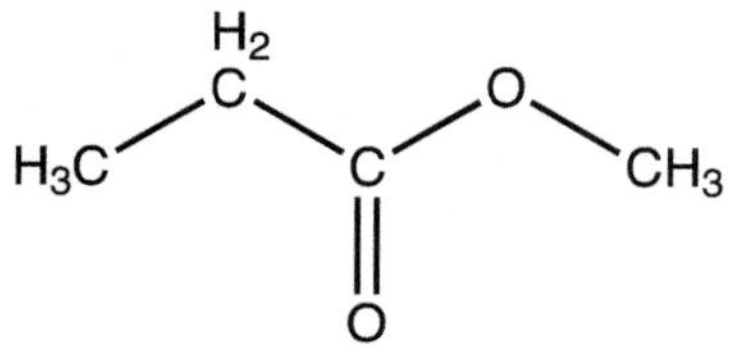

methyl propanoate

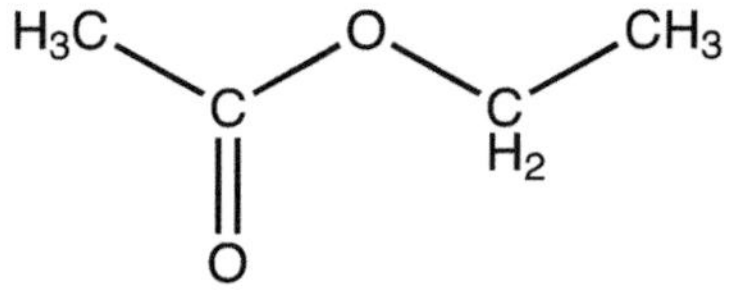

ethyl ethanoate

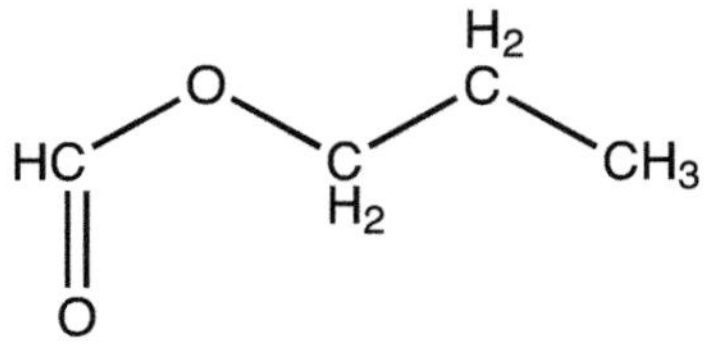

propyl methanoate

isomers of $C_5H_{10}O$

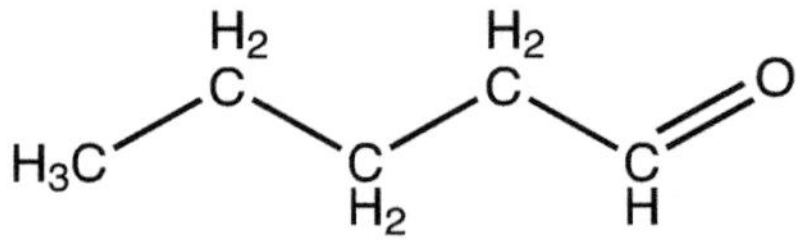

pentanal

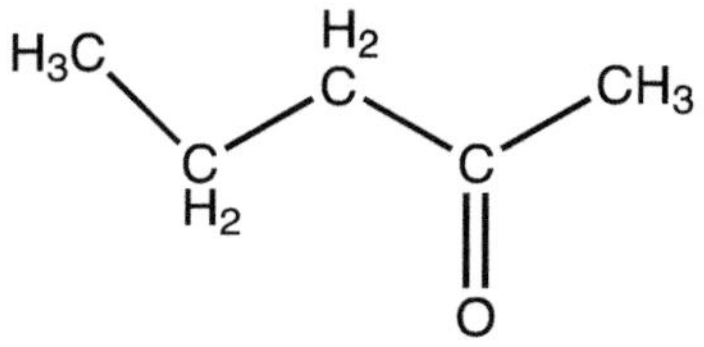

pentan-2-one

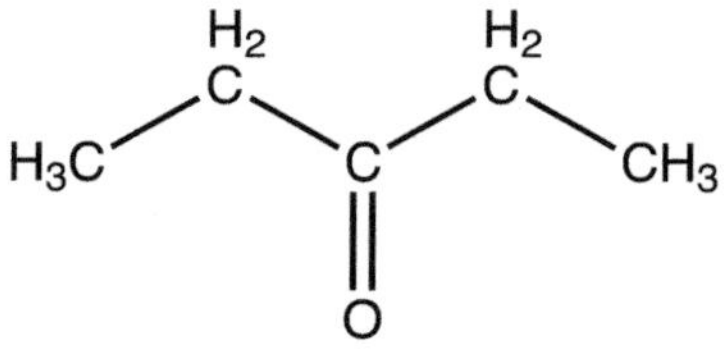

pentan-3-one

Learning Check 10.6

polybut-2-ene — [structure: repeating unit with two CH groups, each bearing H and CH₃]

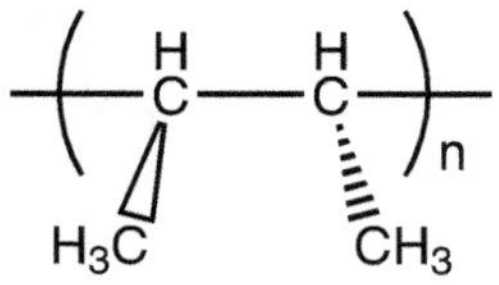

polybut-1-ene — [structure: repeating unit with CH(CH₂CH₃)–CH₂]

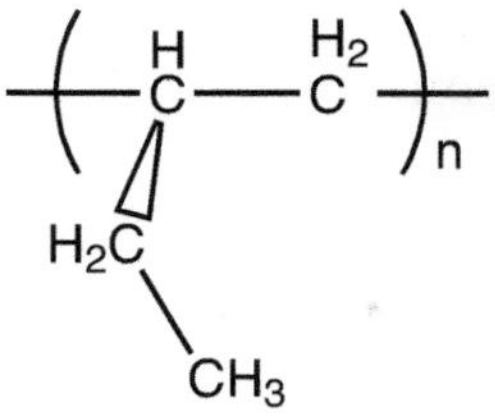

Learning Check 10.7

$2CH_3OH + 3O_2 \rightarrow 2CO_2 + 4H_2O$
$C_2H_5OH + 3O_2 \rightarrow 2CO_2 + 3H_2O$
$2C_3H_7OH + 9O_2 \rightarrow 6CO_2 + 8H_2O$
$C_4H_9OH + 6O_2 \rightarrow 4CO_2 + 5H_2O$

Learning Check 10.8

1-chlorohexane
2-chlorohexane
3-chlorohexane
1-chloro-2-methylpentane
2-chloro-2-methylpentane
3-chloro-2-methylpentane
4-chloro-2-methylpentane (2-chloro-4-methylpentane)
5-chloro-2-methylpentane (1-chloro-4-methypentane)
1-chloro-3-methylpentane
2-chloro-3-methylpentane
3-chloro-3-methylpentane
1-chloro-2,2-dimethylbutane
3-chloro-2,2-dimethylbutane
4-chloro-2,2-dimethylbutane (1-chloro-3,3-dimethylbutane)
1-chloro-2,3-dimethylbutane
2-chloro-2,3-dimethylbutane

Learning Check 10.9

a) Heat but-2-ene with steam and a platinum catalyst to form butan-2-ol. Oxidize butan-2-ol with acidified potassium dichromate to from butanone

b) Treat 1-bromobutane with NaOH to form butan-1-ol. Distill butanal from the reaction of butan-1-ol and acidified potassium dichromate.

Learning Check 10.10

1. React ethene with HBr to form bromoethane. Treat bromoethane with hydrogen cyanide to form propanenitrile

Chapter 10 Summary Questions

1. a) 2-methylbutane

   b) pentane

   c) 2-methylbutane

   d) 3,4-dimethylhexane

2. a) butan-1-ol $H_3C-CH_2-CH_2-CH_2-OH$

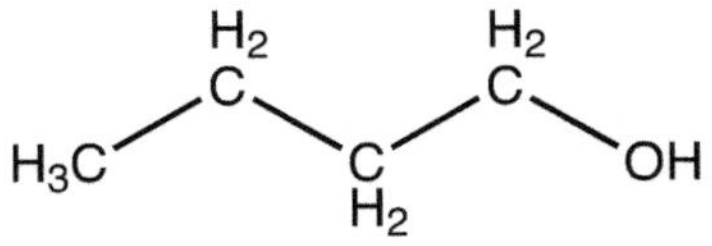

butan-2-ol

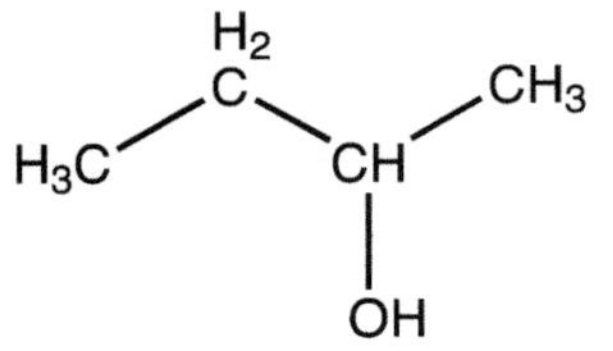

$$H_3C-\underset{H_2}{C}-\underset{\underset{OH}{|}}{CH}-CH_3$$

2-methylbutan-1-ol

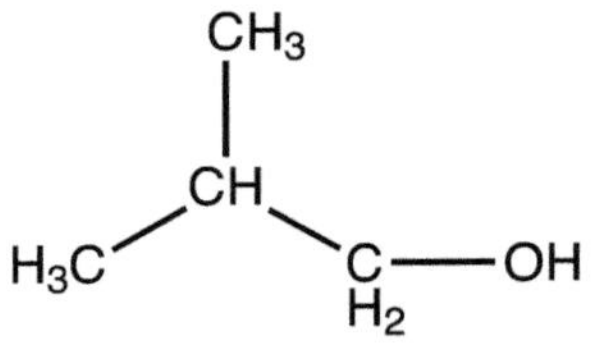

$$H_3C-\underset{}{\overset{CH_3}{\underset{|}{CH}}}-\underset{H_2}{C}-OH$$

2-methylbutan-2-ol

$$H_3C\cdots\underset{\overset{|}{CH_3}}{\overset{CH_3}{\underset{|}{C}}}-OH$$

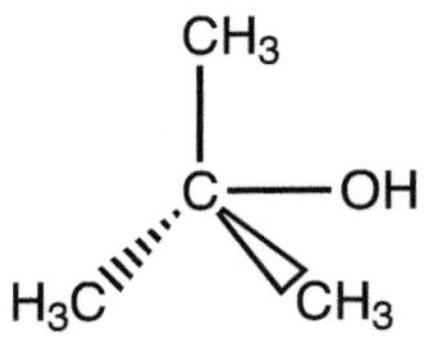

b) i) butanal

   ii) heat with acidified potassium dichromate during distillation.

c) 2-methylpropan-2-ol

d) Heat the primary alcohols with acidified potassium dichromate under reflux.

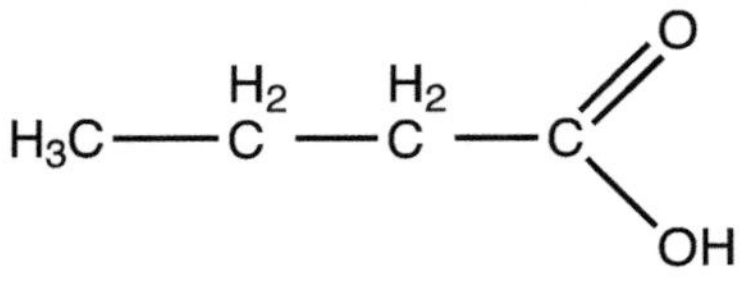

$$H_3C-\underset{H_2}{C}-\underset{H_2}{C}-\underset{OH}{\overset{O}{\overset{\|}{C}}}$$

ii) butanoic acid

$$H_3C-\underset{\underset{CH_3}{|}}{\overset{H}{\underset{|}{C}}}-\underset{OH}{\overset{O}{\overset{\|}{C}}}$$

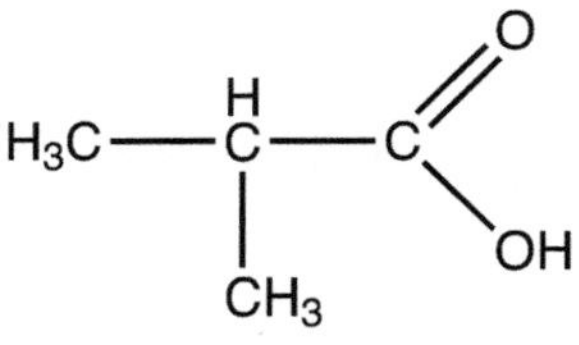

2-methylpropanoic acid

e) i) The reaction occuring when a species with a lone pair of electrons displaces a leaving group from a carbon bearing a partially positive charge.

   ii) C4H9Br + OH- C4H9OH + Br-

   iii) 2-methylpropan-2-ol formed from 2-bromo-2-methylpropane

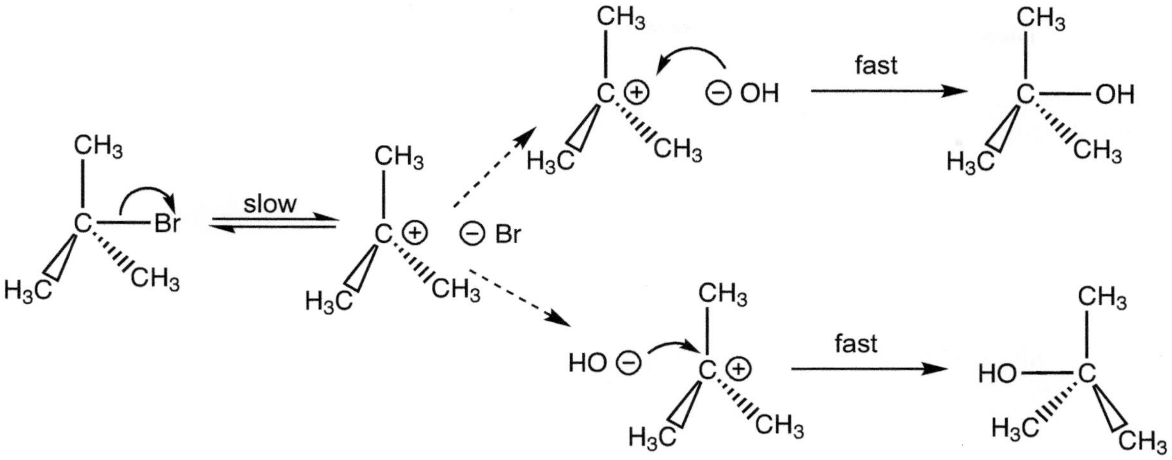

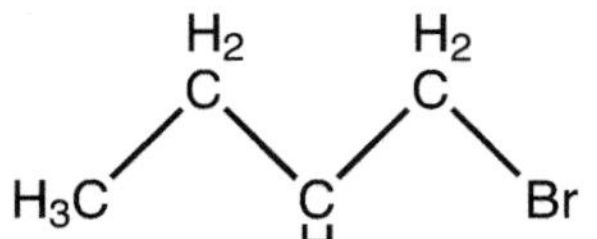

Answers

[Mechanism diagram: (CH3)3C–Br ⇌ (slow) (CH3)3C⁺ + Br⁻, carbocation attacked by ⁻OH from either face (fast) to give (CH3)3C–OH]

iv) 1-bromobutane

[Structure: H3C–CH2–CH2–CH2–Br]

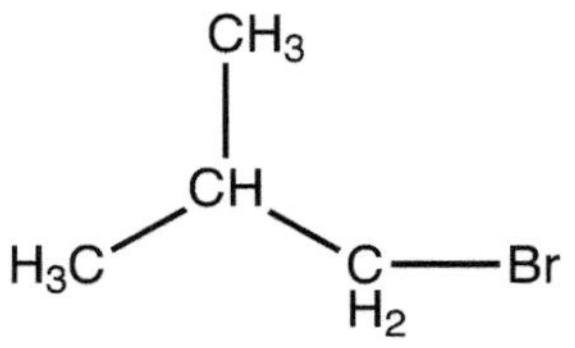

1-bromo-2-methylpropane

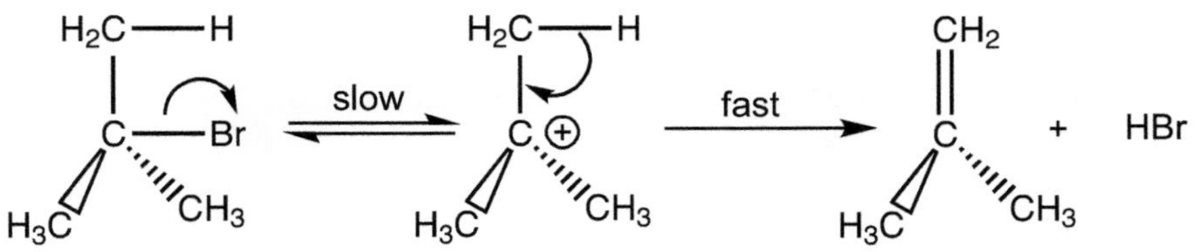

[Structure: (CH3)2CH–CH2–Br]

[Mechanism diagram: (CH3)3C–Br ⇌ (slow) (CH3)3C⁺ → (fast) CH2=C(CH3)2 + HBr]

2-methylpropene

g) butan-2-ol

page 205

Learning Check 11.1

a) 3 s.d.
b) 3
c) 3
d) 4
e) 4
f) 3

Learning Check 11.2

a) 2200
b) 0.035
c) 2.00
d) 0.026
e) 0.00501
f) 2150

Learning Check 11.3

a) 5.87
b) 14.98
c) 0.0275
d) 0.04
e) 842.4
f) 2.7
g) 2.7
h) 12000